Environmental Chemometrics

Principles and
Modern Applications

ANALYTICAL CHEMISTRY

Environmental Chemometrics

*Principles and
Modern Applications*

Grady Hanrahan

CRC Press
Taylor & Francis Group
Boca Raton London New York

CRC Press is an imprint of the
Taylor & Francis Group, an **informa** business

CRC Press
Taylor & Francis Group
6000 Broken Sound Parkway NW, Suite 300
Boca Raton, FL 33487-2742

First issued in paperback 2019

ISBN-13: 978-1-4200-6796-5 (hbk)
ISBN-13: 978-0-367-38634-4 (pbk)

Library of Congress Cataloging-in-Publication Data

Hanrahan, Grady.
    Environmental chemometrics : principles and modern applications / Grady Hanrahan.
        p. cm. -- (Analytical chemistry ; 4)
        Includes bibliographical references and index.
        ISBN 978-1-4200-6796-5 (alk. paper)
        1. Chemometrics. 2. Environmental chemistry. I. Title.

    QD75.4.C45H36 2008
    543.01'5195--dc22   2008029542

Visit the Taylor & Francis Web site at
http://www.taylorandfrancis.com

and the CRC Press Web site at
http://www.crcpress.com

This book is dedicated to my family for their love and support of my educational endeavors. Final dedication goes to Piyawan for her love and patience during the writing process.

Contents

Preface

The heterogeneity of complex environmental systems raises fundamental difficulties with regard to the ability of investigators to detect analytes of interest, extrapolate results, and to correlate variables and predict patterns within samples and differing sample matrices. Fortunately, the area of chemometrics is contributing to a better understanding and mitigation of these difficulties by providing powerful multivariate statistical approaches for design, analysis, and data interpretation.

This book was born out of my frustration when trying to find suitable material with which to educate students, researchers, and laboratory personnel about the rigors of modern environmental analysis, and how to solve limitations effectively through multivariate approaches. The intent of this book is to introduce modern chemometric methods, the basic theory behind them, and their incorporation into current environmental investigations. It is written in a way that limits the need for a background in advanced statistical methods, with the inclusion of real-world applications to strengthen learning and comprehension. All major areas of environmental analysis are covered, including discussions of the fundamental principles underlying chemometric techniques backed by studies from experts in the field.

A feature of this book is the inclusion of expanded research applications at the end of each chapter, where detailed examinations of published studies are presented with special emphasis on how chemometric methods are appropriately incorporated. Also included are introductions to matrix calculations and the use of Excel spreadsheet functions, with examples of Excel exercises provided in Chapter 2. This book is by no means comprehensive, but it does provide key content and applications in all the major areas of chemometric analysis. The use of a solely dedicated software package was avoided as investigators have numerous options with varying degrees of complexity from which to choose. Note, however, that many figures displayed in the simulated applications were performed by the use of JMP (SAS Institute, Inc.) software. Ultimately, it is expected that this book will be highly useful to a variety of readers, especially those intimately involved in environmental analysis.

Grady Hanrahan

Acknowledgments

I am grateful to a number of people involved in the preparation of this book. I must first thank Lindsey Hoffmeister, Fiona Macdonald, and Dr. Charles Lochmüller for believing in the concept of the book and for providing valuable direction during the beginning stages of its development. I am also grateful to Hilary Rowe, David Fausel, and Tom Skipp of the Taylor & Francis Group, LLC, for their support during all stages of development. I thank the blind reviewers for providing valuable comments and direction, and Drs. Krishna Foster, Scott Nickolaisen, Ira Goldberg, William Bilodeau, and Sally Bilodeau for reviewing individual chapters. They have made this a more complete piece of work. I also thank those investigators who have provided helpful comments and original figures of the relevant applications housed within individual chapters. Special thanks also goes out to Vicki Wright, Kanjana Patcharaprasertsook, and Katherine Snyder for their valuable help with word processing and figure development activities. Finally, I owe a debt of gratitude to all my former and current students who have encouraged me to become a better educator.

Author

Grady Hanrahan is the Stauffer Endowed Chair of Analytical Chemistry at California Lutheran University. He received his PhD training in environmental analytical chemistry from the University of Plymouth, UK. With experience in directing undergraduate and graduate research, he has taught in the fields of environmental science and analytical chemistry at the California State University, Los Angeles. He is the author or coauthor of over 30 technical papers and is active in employing chemometric methods in his research efforts.

1 Introduction

1.1 CHEMOMETRICS—AN OVERVIEW

What is *chemometrics*? In its broadest definition, chemometrics is a subdiscipline of analytical chemistry that uses mathematical, statistical, and formal logic to design or select optimal experimental procedures, to provide maximum relevant chemical information by analyzing chemical data, and to obtain knowledge about given chemical systems (Massart et al. 1997).

Although statistical methodologies such as *curve fitting* and *statistical control* were used in analytical chemistry throughout the 1960s (Currie et al. 1972), it was not until 1971 that Svante Wold coined the term *chemometrics* (Brereton 1990, Hopke 2003). The broad definition used here has been shaped by the evolution of this subdiscipline over the past 35 years. It was Bruce Kowalski, in collaboration with Wold, who introduced the term *chemometrics* into the United States (Hopke 2003). In June 1974, they established the International Chemometrics Society. Today, there are a variety of Chapters of the International Chemometrics Society that provide resources for professionals working in the field, and a forum for the exchange of ideas. The first known paper with chemometrics in the title, which presented the value of pattern recognition concepts to the analytical community, was subsequently written by Kowalski in 1975.

The 1980s brought about an era of enhanced computing capabilities and increasingly sophisticated analytical instrumentation. The torrent of data generated by these multielement and multicomponent instruments required the application of already-established chemometric methods as well as creating a need for higher-level methodologies. Such methods were presented to the scientific community with the advent of two specialized journals: *Chemometrics and Intelligent Laboratory Systems*, established in 1986, and *Journal of Chemometrics*, in 1987.

An increased number of investigators began incorporating chemometrics into their research activities in the 1990s. Brown et al. (1996) reported approximately 25,000 computer-generated citations in a comprehensive 1996 review of chemometrics. Wold and Sjöström presented an informative look at the success of chemometrics in a 1998 review (Wold and Sjöström 1998). They illustrated how chemistry is driven by chemometrics and described state-of-the-art chemometric methods that included multivariate calibration, structure–(re)activity modeling and pattern recognition, and classification and discriminant analysis.

> "It is evident that chemometric techniques are, and will continue to have, a profound effect on other research areas including... environmental science."

The twenty-first century has brought about even greater sophistication with an increasing trend toward miniaturized analytical devices (e.g., lab-on-chip) allowing automated, high throughput capabilities with low reagent and sample use. In a more recent review, Lavine and Workman (2004) described the use of chemometrics in such areas as image analysis, sensors, and microarrays. They described a new paradigm of "cause-and-effect information discovery from data streams." Considering the information presented, chemometric methodologies are increasingly important tools in modern analytical chemistry. Chemometric techniques will therefore have a profound effect on other research areas including chemical engineering, drug design, food technology, biomedical research, and environmental science.

1.2   THE IMPORTANCE OF QUANTITATIVE ENVIRONMENTAL ANALYSIS

The global society is currently experiencing a high degree of industrialization that will ultimately bring about improvements in the physical environment and general well-being of mankind. Unfortunately, this acceleration is resulting in a burgeoning population, increasing pollution, and significant local and global environmental catastrophes. Thus, to maintain a healthy environment, environmental protection measures are being continuously increased through the establishment of stringent laws, policies, and regulations.

The success of environmental protection relies on a multitude of parameters including proper implementation of policies, project management, environmental technology development, and effective monitoring programs. Monitoring of potential pollutants, and of the overall quality of the environment, is generally based upon the quantitative analysis of abiotic matrices such as air, soil, sediment, and water. The heterogeneity of these differing, and often complex, matrices raises difficulties with regard to the ability of investigators to detect potential pollutants of interest and to extrapolate analytical results. The implementation of a quantitative environmental analysis program allows for the proper identification of pollutants and determines their fate in the environment. The strength of such a program is based on the accurate and sensitive determination of pollutants, accompanied by sound *quality assurance* (QA) and *quality control* (QC) measures to limit errors and discrepancies during the analysis process. Further detailed discussions of QA and QC measures are presented in Chapter 3. The importance of quantitative environmental analysis lies in the effective and successful implementation of environmental pollution strategies.

1.3   COMMON CHEMICAL AND BIOLOGICAL POLLUTANTS IN ENVIRONMENTAL MATRICES

In broad terms, a *pollutant* can be defined as a human-generated or naturally occurring chemical, physical, or biological agent that adversely affects the health of humans and other living organisms and/or modifies the natural characteristics of a given ecosystem. It is beyond the scope of this book to cover all potential pollutants in each environmental matrix. Discussion is therefore centered on common human-generated pollutants, while acknowledging that natural occurrences (e.g., volcanic eruptions) can contribute significantly to pollution.

Pollutants can arise from a single identifiable source (*point source*) such as industrial discharge pipes or factory smokestacks. They can also be generated from a source that is much harder to identify and control (*nonpoint source*). Examples of nonpoint sources include land runoff and atmospheric deposition. For further clarity, expanded discussions on each environmental matrix and potential pollutant source are presented in Sections 1.3.1 to 1.3.3. Pollutant transport and transformation processes that occur among the different matrices (e.g., sediment–water exchange, surface runoff, atmospheric deposition onto land and water bodies) are also covered in generalized discussions. Detailed discussions on relative concentrations, along with the significance and fate of pollutants in these systems, are given in application-based sections throughout this book.

1.3.1 AIR

Air pollution is a global problem caused by the emission of toxic pollutants, which either alone or through various chemical reactions, leads to adverse health and environmental impacts. Accordingly, many nations have established stringent air quality standards to try to reduce the emission of potential pollutants. The US Clean Air Act Amendments of 1990, for example, established a list of 189 toxic air pollutants that must have their emissions reduced (Novello and Martineau 2005). This list contains source categories that include (a) small sources (e.g., dry cleaning facilities) and (b) major sources emitting 10 tons per year of any pollutant, or 25 tons per year of any combination of those pollutants (Cheremisinoff 2002). There have been similar interim recommendations by the European Union (EU) Communities 6th Environmental Action Program's Thematic Strategy on Air Pollution (Commission of the European Communities 2005). A commonality between both is the concentration on a set of criteria pollutants, those identified originally by the US Clean Air Act of 1970 as having the greatest health and environmentally-related threat. Table 1.1 lists the original six *criteria air pollutants* for which National Ambient Air Quality Standards were set to protect public health (*primary standards*) and public welfare including damage to animals, crops, vegetation, and buildings (*secondary standards*). Primary standards [as set by the U.S. Environmental Protection Agency (EPA)] and lower assessment thresholds (as recommended by the EU Thematic Strategy on Air Pollution) to protect human health are also included. In addition, volatile organic compounds (VOCs) are also discussed in this chapter, which are emitted in large amounts and are often associated with criteria pollutants (Hill 2004). They are emitted as a gas from both solids and liquids, and include such sources as burning fuel, paints and lacquers, building supplies, cleaning products, glue, office equipment, and correction fluids. Both chemical and physical characteristics that contribute to their toxic effects are described for each pollutant listed.

Of the six criteria pollutants, ground-level ozone (O_3) and particulate matter are considered to pose the greatest health threats (Kongtip et al. 2006). *Ozone* is a

TABLE 1.1

Criteria Air Pollutants with Ambient Air Quality Standards/Thresholds and Chemical and Physical Characteristics

Air Pollutants	Primary Standards (U.S. EPA)[a,b]	Lower Assessment Thresholds (EU)[a,c]	Chemical and Physical Characteristics
Carbon monoxide (CO)	10 mg/m³ (8 hour average)	5 mg/m³ (8 hour average)	Colorless, tasteless and odorless gas. Formed as a result of combustion processes (e.g., industrial, motor vehicle exhaust).
Lead (Pb)	1.5 µg/m³ (quarterly average)	0.25 mg/m³ (annual average)	Toxic metal with high mobility and soluble in acid environments.
Nitrogen oxides (NO$_x$)[d]	100 µg/m³ (annual average)	26 µg/m³ (annual limit value)	Consist primarily of nitric oxide (NO), nitrogen dioxide (NO$_2$), and nitrous oxide (N$_2$O). NO$_x$ are precursors to both acid precipitation and ground-level ozone.
Ozone (O$_3$)	0.08 ppm (8 hour average)	N/A[e]	Colorless, reactive oxidant gas and a major constituent of atmospheric smog. Formed by the photochemical reaction of sunlight, VOCs, and NO$_x$.
Particulate matter (PM$_{2.5}$)	15 µg/m³ (annual average)	7 µg/m³ (annual average)	Complex mixture of fine organic and inorganic substances. Sulfate ions contribute the
Particulate matter (PM$_{10}$)	150 µg/m³ (24 hour)[f]	20 µg/m³ (24 hour)	greatest portion of fine particulate matter. Smaller particles contain secondary formed aerosols, combustion particles, and organic and metal vapors.
Sulfur oxides (SO$_x$)	0.14 ppm (24 hour)[g]	50 µg/m³ (24 hour)[h]	Compounds formed of sulfur and oxygen molecules with sulfur dioxide (SO$_2$) being the predominant form.

[a] Units as reported.
[b] U.S. Clean Air Act, 1990.
[c] As recommended by the EU Thematic Strategy on Air Pollution.
[d] Standards reported for nitrogen dioxide (NO$_2$).
[e] Importance of reduction stressed, but no threshold set.
[f] Due to the lack of evidence linking health problems to long-term exposure to PM$_{10}$, the U.S. EPA revoked the national ambient air quality standard in 2006.
[g] Not to be exceeded more than once in a given calendar year.
[h] Reported as sulfur dioxide (SO$_2$)—not be exceeded more than three times in a given calendar year.

harmful respiratory irritant that can constrict airways and interfere with the lung's immunity. It is created by the chemical reaction between nitrogen oxides (NO_x) and VOCs (e.g., anthropogenic aromatics, alkenes) in the presence of sunlight and is a major constituent of photochemical smog. It is not surprising then, that the city of Los Angeles was at the top of the list in this category for the most polluted city in the United States in 2006 (American Lung Association 2006). The deadly cocktail of hot, stagnant air, sunny days and concentrated vehicular traffic makes it an ideal environment for ground-level ozone formation. Table 1.2 provides information on the number of days per year that the Los Angeles Basin exceeded the health standard levels of permitted air pollution between 1995 and 2005.

The good, the bad, and the ugly

Do not confuse "good" ozone—that occurring in the stratosphere (approximately 10–30 miles above the Earth) to protect life from the sun's harmful rays—with "bad" ozone—the ground-level constituent of photochemical smog layers found in surrounding environs of many urban cities.

Conversely, ozone production is more limited when cloudy, cool, rainy, and windy conditions exist. Statistical models that account for weather-related variability are thus used as a more accurate assessment of underlying ozone trends. Fortunately, the present trend is one of decreasing ozone levels across the major cities of the world due to the mandatory reduction of NO_x and hydrocarbons containing $C{=}C$ bonds plus other reactive VOCs. This trend is evident in the data presented in Table 1.2.

TABLE 1.2

Number of Los Angeles Basin Calendar Days Exceeding Health Standard Levels Established for Ozone

Year	State Ozone Standard (1 hour average > 0.12 ppm)	Federal Ozone Standard (8 hour average > 0.08 ppm)
1995	154	127
1996	151	132
1997	141	127
1998	114	111
1999	118	113
2000	123	111
2001	121	100
2002	118	99
2003	133	120
2004	111	90
2005	102	84

Source: http://www.aqmd.gov/hb/govbd.html.

Particulate matter (PM) is a complex mixture of dust, pollen, acids, organic chemicals, metals, and soil particles, and the size of the particles is directly linked to harmful health effects (Pyne 2002). Typically, particles with a diameter smaller than 2.5 μm ($PM_{2.5}$), termed fine particles, are inhaled deeper into the lung tissue, and are therefore more harmful. Particles larger than 2.5 μm and smaller than 10 μm (PM_{10}) are termed inhalable coarse particles. However, due to the lack of evidence linking health problems with long-term exposure to PM_{10}, the U.S. EPA revoked the Annual National Ambient Air Quality Standard in 2006. The PM_{10} particles, however, are still studied heavily across the globe (see Expanded Research Application V in Section 6.12 of this book for multivariate studies of PM_{10} components in relation to cell toxicity).

There are also other critical pollutants involved in poor air quality, including carbon monoxide (CO), lead (Pb), and sulfur oxides (SO_x). Carbon monoxide is estimated to account for more than 50% of air pollution worldwide (Hill 2004), and is the result of incomplete combustion of carbon-based material (e.g., fossil fuel and biomass). This must not be confused with carbon dioxide (CO_2), which is a product of complete combustion and is much less toxic. See Figure 1.1 for the structural differences. The structure of the CO molecule is best described using the molecular orbital theory described in general chemistry textbooks. The length of the bond (0.111 nm) indicates that it has a partial triple bond character as depicted in Figure 1.1b.

The amount of lead has decreased over recent years because of the phasing out of leaded gasoline. However, stationary sources such as lead smelters, incinerators, utilities, and lead acid battery manufacturers can produce toxic levels in isolated areas. Lead exposure particularly affects young children and deposits in soil and water can harm animals and fish. Sulfur oxides, in particular sulfur dioxide (SO_2), are strong respiratory irritants generated by the combustion of gasoline, power plants, and other industrial facilities. Sulfur dioxide contributes to the formation of PM that can be transported over long distances by wind patterns and contributes to low visibility haze. Oxidation is most common in clouds and especially in heavily polluted air where compounds such as ammonia and ground-level ozone abound. Sulfur dioxide is converted into sulfuric acid (H_2SO_4) under wet (e.g., rainy) conditions (see Example Problems 1.1).

Under dry conditions [in the presence of atmospheric oxygen (O_2)], sulfur dioxide is converted into sulfate (SO_4). Ostensibly, atmospheric compounds like SO_2 can exchange all along the Earth's surface under dry conditions, including dissolution

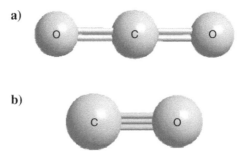

FIGURE 1.1   Structural characteristics of (a) carbon dioxide and (b) carbon monoxide.

and reaction with the sea surface, with the aqueous phase of living organisms, or with basic soil types (Butcher et al. 1992). Horizontal and vertical transport processes coupled with chemical reactions in the atmosphere are also involved in the transport and diffusion of other trace substances [e.g., the transport of iron (Fe) in windblown soils].

EXAMPLE PROBLEMS 1.1

1. Give logical reasoning why "incomplete" combustion is likely to result in the formation of excess carbon monoxide (CO) rather than carbon dioxide (CO_2).

Answer: Only under optimal burning conditions, with excess oxygen, can carbon be completely oxidized to CO_2.

2. Provide a detailed chemical mechanism by which sulfur dioxide (SO_2) is converted to sulfuric acid (H_2SO_4), ultimately contributing to acid rain.

Answer: SO_2 reacts with oxygen in the atmosphere to form sulfur trioxide (SO_3), which then reacts readily with H_2O to form H_2SO_4. The mechanism involves hydroxyl radicals and can be shown in complete sequence as follows:

$$SO_2 + OH^{\bullet} \longrightarrow HSO_3^{\bullet}$$

$$HSO_3^{\bullet} + O_2 \longrightarrow SO_3 + HOO^{\bullet}$$

$$SO_3 + H_2O \longrightarrow H_2SO_4 \text{ (g)}$$

$$H_2SO_4 + H_2O \longrightarrow H_2SO_4 \text{ (aq)}$$

Note: Most of the oxidation of SO_2 to H_2SO_4 occurs in the aqueous (aq) phase.

1.3.2 WATER

Water is arguably one of the most precious and ubiquitous substances on Earth. It is a resource that is often taken for granted by most people who give little thought to the processes involved to purify it before it comes out of the tap. Water of this quality is termed *potable*—defined by the U.S. EPA as "water which meets the quality standards prescribed in the U.S. Public Health Service Drinking Water Standards, or water which is approved for drinking purposes by the State or local authority having jurisdiction" (U.S. EPA 1994). High quality potable water from where drinking water is obtained can also be used in essential industrial applications (e.g., food and beverage production), textile manufacturing, livestock watering, irrigation, and in the microelectronics industry (Worsfold et al. 2008). The impact of such water on humans is governed with guidelines established by such authorities as the U.S. EPA, the World Health Organization (WHO), the EU Community, and the Australian National Health and Medical Research Council. Table 1.3 provides WHO guidelines for selected drinking water parameters that are implied in the term *guidelines*, based on scientific research and epidemiological findings (WHO 2006).

TABLE 1.3
WHO Guidelines for Selected Drinking Water Quality Criteria/Parameters

Water Quality Criteria/Parameter	Guideline Value[a]
Arsenic	0.005
Barium	0.700
Cadmium	0.003
Color	15 TCU[b]
Cadmium	0.003
Dichlorobenzene	5–50
E. coli or thermo-tolerant Coliform bacteria	Not detectable in any 100 mL sample for water intended for human consumption
Hardness, pH, DO	Not specified
Iron	0.3
Lead	0.010
Mercury	0.001
Nitrate	50
Selenium	0.010
Styrene	4–2600
Sulfate	250
Toluene	24–170
Total Dissolved Solids (TDS)	1000
Turbidity	5 NTU[c]
Zinc	3

[a] All values in mg/L unless otherwise stated.
[b] One true color unit (TCU), or platinum–cobalt unit, corresponds to the amount of color exhibited under the specified test conditions by a standard solution containing 1.0 mg/L of platinum.
[c] Turbidity is measured in nephelometric turbidity units (NTU), ostensibly how much light reflects for a given amount of particulate matter.

Society must consider which indicators are important for the establishment of water quality guidelines for ecosystem protection. Fresh and marine water systems, including rivers, lakes, streams, wetlands, oceans, and coastal waterways, are vital to the balance of life on Earth and every effort must be made to protect them. Here, physicochemical indicators dominate, but an increasing awareness of biological stressors (e.g., species richness, primary production) has emerged in the last decade.

Physicochemical indicators can be classified as those that have direct toxic effects on the biota (e.g., heavy metals, temperature) and those that affect ecosystems directly (e.g., nutrients, turbidity, excess organic matter) (Hart 2002). Excess *nutrients* (e.g., phosphorus, nitrogen) in slow-moving water bodies, for example, can lead to *eutrophication*—a process of excess plant growth (e.g., algae blooms), ultimately reducing the amount of *dissolved oxygen* (DO) in the water and killing off vital organisms (Hanrahan et al. 2001). Thus, assessing nutrient concentrations on a regular basis is of paramount importance for providing insight into the relative health of aqueous environments.

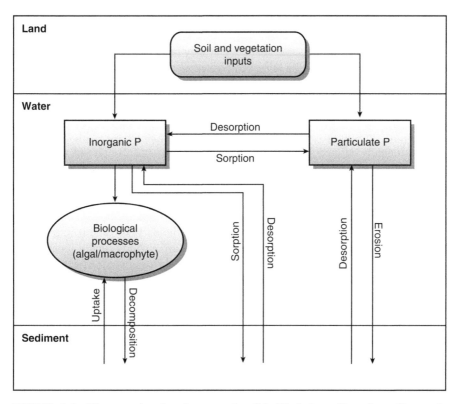

FIGURE 1.2 The aquatic phosphorus cycle. (Modified from Hanrahan, G., et al., *Environmental Monitoring Handbook*, McGraw-Hill, New York, 2002. With permission from Elsevier.)

The term *physicochemical* is often associated with environmental analysis and refers to both physical and chemical properties of a given system.

Figure 1.2 shows the distribution and transformation of phosphorus (P) in aquatic systems. It provides an example of the processes that can distribute and transform a species between matrices (soil, water, and sediments). Unlike nitrogen, phosphorus does not have a significant atmospheric component. Chemical *speciation* of phosphorus between dissolved and particulate components does occur via, for example, the adsorption and precipitation processes. In addition, biological processes (e.g., algal) are involved in uptake and decomposition.

The term *speciation* concerns the identification and quantitation of specific forms of elements traditionally given by operationally defined results to identify such forms as "bioavailable"—those readily used by biological organisms.

Operationally defined species are defined by the methods used to separate them from other forms of the same element that may be present. Nutrient and metal species, for example, can be separated between "soluble" and "insoluble" forms by passing the aqueous sample through a 0.20 or 0.45 μM membrane filter.

Overall, phosphorus does occur in both particulate and dissolved forms and can be operationally defined as total phosphorus (TP), total reactive phosphorus (TRP), filterable reactive phosphorus (FRP), and total filterable phosphorus (TFP) (Hanrahan et al. 2002). TP is a measure of both dissolved and particulate P forms and has been proposed as a useful indicator of mean chlorophyll concentrations in streams (Van Nieuwenhuyse and Jones 1996). FRP, however, is the *fraction* considered most readily available for algal and macrophyte growth (Reynolds 1984), and is important in relation to the concept of eutrophication discussed earlier. Additionally, recent developments in the field of microbiology and research on the origin of life have suggested a possibly significant role for reduced, inorganic forms of phosphorus in bacterial metabolism and as evolutionary precursors of biological phosphate compounds (Hanrahan et al. 2005). Reduced inorganic forms of phosphorus include phosphorus acid [H_3PO_3, P(+III)], hypophosphorus acid [H_3PO_2, P(+I)], and various forms of phosphides [P(−III)]. Such compounds have been detected in anaerobic sediments, sewage treatment facilities, and in industrial and agricultural processes. Figure 1.3 depicts the calculated equilibrium distributions of the various species of phosphate, hypophosphite, and phosphite as a function of pH. At the circumneutral pH of most natural waters and soils, the dominant P species according to equilibrium calculations are: $H_2PO_4^-$ and HPO_4^{2-} for phosphate; $H_2PO_3^-$ and HPO_3^{2-} for phosphite; and $H_2PO_2^-$ for hypophosphite. The charge of each species will determine the environmentally relevant reactions (such as sorption/desorption, Figure 1.2) that may influence its mobility and distribution. Furthermore, the level of protonation of a given chemical species will influence its detection.

Like nutrients, *metals* are present in a variety of forms in aqueous systems with typical fractions measured as presented in Figure 1.4. The determination of individual metal species is often necessary as total concentrations are poor indicators to understanding their availability and toxicity to organisms and overall mobility in natural systems (Donat and Bruland 1995). Typically, only a relatively small fraction of the total dissolved metal concentration is present in the free hydrated state or complexed with inorganic ligands, while a large fraction is complexed with organic ligands (Apte et al. 2002). Acid-extractable metals are determined after the unfiltered sample is treated with hot dilute mineral acid prior to analysis. Factors influencing the speciation of metals include pH, temperature, alkalinity, major ion composition (e.g., metal-carbonate complexes), and ionic strength.

At present, there is no universal technique for measuring all metal species. Nonetheless, there are a number of analytical methodologies available to measure, for example, the filterable portion that is linked to metal bioavailability. Such methods include (but are not limited to) solvent extraction atomic absorption spectrometry, anodic stripping voltammetry (ASV), cathodic stripping voltammetry (CSV), capillary

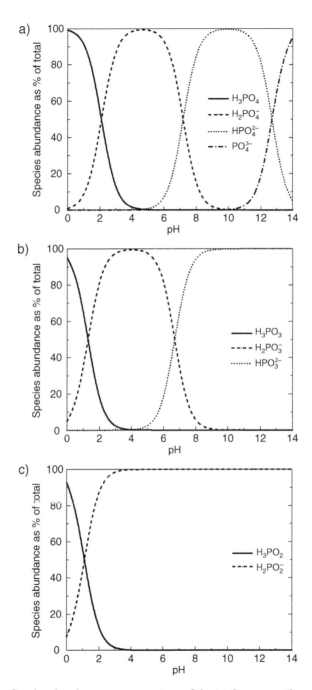

FIGURE 1.3 Species abundance, as a percentage of the total concentration, are shown for (a) phosphate, (b) phosphite, and (c) hypophosphite species. At circumneutral pH (typical of surface waters), the dominant species are $H_2PO_4^-$ and HPO_4^{2-} for phosphate, $H_2PO_3^-$ and HPO_3^{2-} for phosphite, and $H_2PO_2^-$ for hypophosphite. (From Hanrahan, G., et al., *Talanta*, 66, 435–444, 2005. With permission from Elsevier.)

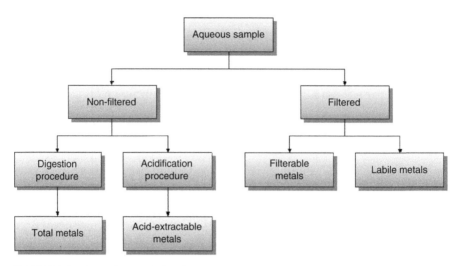

FIGURE 1.4 Operationally defined metal fractions in aqueous environments. (Adapted from Apte, S. C., Batley, G., and Maher, W. A., *Environmental Monitoring Handbook*, McGraw-Hill, New York, 2002. With permission from Elsevier.)

electrophoresis (CE), and flow injection analysis (FIA)-related technologies. More detailed discussions on these techniques, along with others to determine metal concentrations, are presented further throughout this book.

A variety of other pollutants and indicators are of importance in aqueous environments. Organic pollutants are significant because of their potential toxicity and carcinogenicity, and include such chemicals as pesticides from agricultural sources, industrial polychlorinated compounds, polynuclear aromatic hydrocarbons (PAHs)

> The term *labile* in Figure 1.2 refers to a relatively unstable and transient chemical species or (less commonly) a relatively stable but reactive species. New analytical techniques such as diffusion gradients in thin films (DGT) are designed to measure *in situ* labile trace metals. For DGT, the metal species from the water column are trapped in a resin gel comprising a Chelex resin after their diffusion across a diffusion layer consisting of a polyacrylamide hydrogel. [See Zhang and Davison (2001) for expanded discussions on DGT.]

and phenols from combustion and petrochemical production, and halogenated compounds derived from water treatment (Worsfold et al. 2008). The International Union of Pure and Applied Chemistry (IUPAC) defines the simplest PAHs as being phenanthrene and anthracene. Smaller and possibly more common to the general public are benzene and naphthalene (see Chapter 3). Although not formally PAHs, they are chemically related to PAHs, and characterized as one-ring (mono) and two-ring (di) aromatics. PAHs may contain four-, five-, six-, or even seven-member rings, but those with five or six are the most common. Figure 1.5 depicts some of the more common PAHs and their structural characteristics.

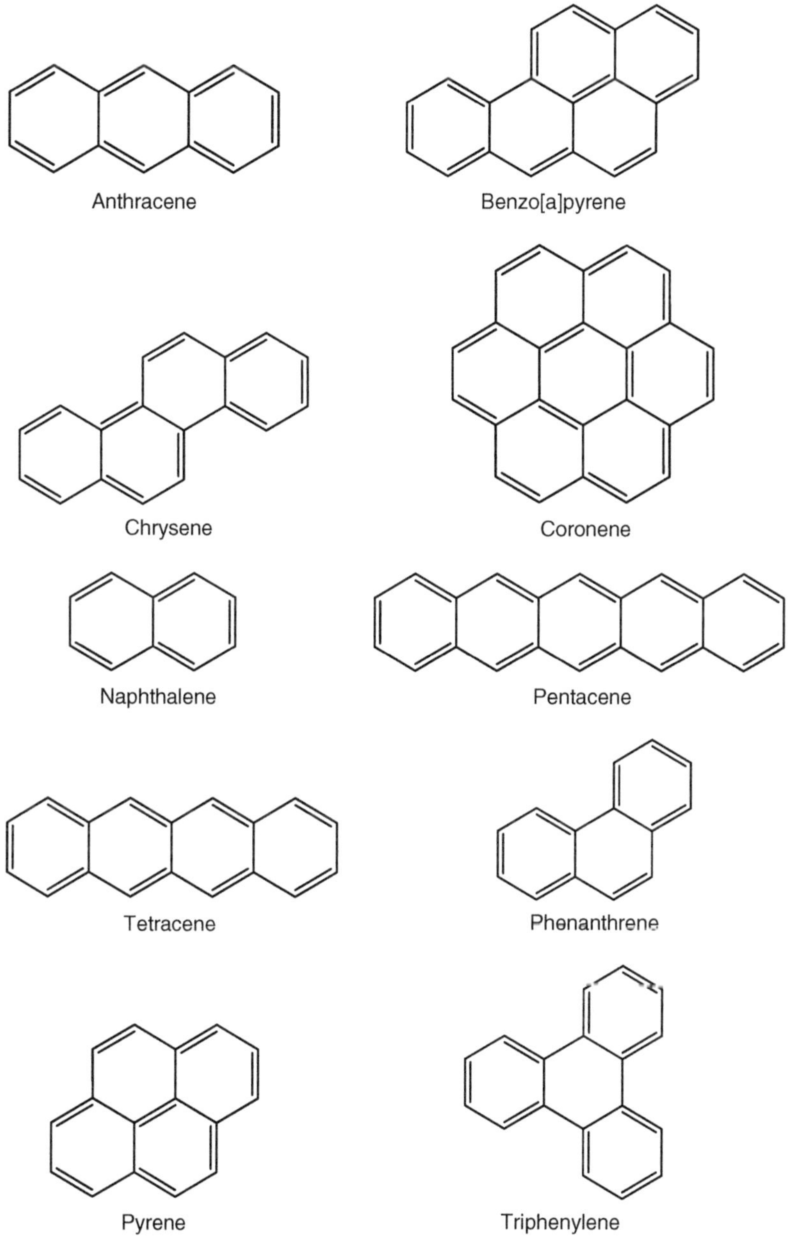

Anthracene

Benzo[a]pyrene

Chrysene

Coronene

Naphthalene

Pentacene

Tetracene

Phenanthrene

Pyrene

Triphenylene

FIGURE 1.5   Common PAHs and their structural characteristics.

High organic matter discharges from point sources can cause oxygen depletion and result in fish and invertebrate population reduction. For this reason, DO concentration is often used as a primary indicator of stream conditions, along with parameters such as dissolved organic carbon (DOC) concentration, biochemical oxygen demand

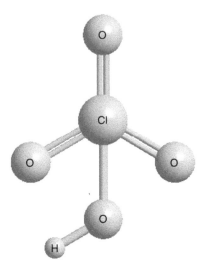

FIGURE 1.6   Structure of perchloric acid.

(BOD), and chemical oxygen demand (COD). Finally, the presence and effect of endocrine-disrupting substances in water is of increasing concern. Such substances include natural hormones, synthetic steroids, and alkyl phenolic surfactants in domestic wastewaters, phytoestrogens in effluent from papermaking, perchlorate from jet fuel and lubricating oils, and phthalates from the plastics industries (Worsfold et al. 2008). These substances are believed to interfere with normal bodily hormone functions, affect the reproduction process, and increase individual susceptibility to cancer and other diseases, even at low parts per billion (ppb) levels (Sullivan et al. 2005). Perchlorates, in particular, have received a lot of press, especially in the Southern California region. Perchlorates are the salts derived from perchloric acid ($HClO_4$, Figure 1.6). They are known to interfere with iodine uptake into the thyroid gland, thus changing thyroid hormone levels that are suspected to cause thyroid gland tumors and improper development and regulation of metabolism in children (Blount et al. 2006).

Dissolved oxygen (DO) is the amount of oxygen, in milligrams per liter, freely available in water and necessary for aquatic life and the oxidation of organic materials.

Biochemical oxygen demand (BOD) is a measure of the quantity of dissolved oxygen, in milligrams per liter, necessary for the decomposition of organic matter (e.g., decomposed plant matter) by microorganisms.

Chemical oxygen demand (COD) is a measure of the chemically oxidizable material in the water and provides an approximation of the amount of organic material present.

In a 2003 memorandum released by the California–Los Angeles Regional Quality Control Board, perchlorate was shown to be present in 38 water supply systems within the Los Angeles Region. The State of California has recognized the severity of the problem and has proposed an action level of 4 ppb. The U.S. EPA is currently in the process of performing toxicity assessments that reflect the science and health effects of perchlorate, an action brought about by the severity of the California occurrences and the additional releases of perchlorate reported in 20 other states across the U.S.A. Some form of regulation is expected once the toxicity assessments are completed.

EXAMPLE PROBLEMS 1.2

1. Why is natural rainwater slightly acidic, with a pH of approximately 5.7?

Answer: Natural rainwater interacts with CO_2 in the atmosphere, forming carbonic acid (H_2CO_3). Some of the carbonic acid in the rainwater then dissociates, producing more hydrogen ions and bicarbonate ions, both of which are dissolved in the rainwater:

$$H_2O + CO_2 \rightarrow H_2CO_3$$

$$H_2CO_3 \rightarrow HCO_3^- + H^+$$

The H^+ ion produced by the second reaction lowers the pH depending on how much HCO_3^- is present in the first reaction. The Earth's atmosphere contains, on average, approximately 0.3% CO_2. Using this value in the two reactions above, one can calculate that the concentration of H^+ in rainwater at a chemical equilibrium of $10^{-5.7}$ moles/L, which is equivalent to a pH of 5.7.

2. As biochemical oxygen demand (BOD) increases in aquatic systems, dissolved oxygen (DO)_____.
   a. increases
   b. decreases
   c. remains the same
   d. none of the above

Answer: b. When BOD levels are high, DO levels decrease because organisms (e.g., bacteria) consume the oxygen that is available in the water.

Many of the factors that influence surface water also influence groundwater composition. Groundwater is in continuous contact with rocks and minerals, and moves more slowly than surface water (centimeters per day instead of kilometers per hour). Once pollutants reach the slow-moving groundwater from the surface, they tend to form a concentrated plume that flows along the *aquifer*. The movement of a pollutant through groundwater is often referred to as pollutant (or contaminant) transport, and the process by which it takes place is called the transport mechanism (e.g., advection, diffusion, and dispersion).

Get your units straight!

Note: ppb is expressed as micrograms of a chemical per liter of water (μg/L). To get an idea of scale, consider that 1 ppb is roughly equal to dissolving one fast-food sized packet of salt into an Olympic-size pool.

Despite the slow movement of pollutants through an aquifer, groundwater pollution can spread over a large area since it can go undetected for many years (Boulding 1995). This is alarming since many of the world's inhabitants rely on groundwater as their major drinking water source. Remediation or clean-up can be difficult due to the complex mix of hydrogeology and chemical considerations. Pollutant diffusion into low-permeability zones, for example, can limit the effectiveness of the pump-and-treat remediation measures that are routinely employed.

An *aquifer* is a subsurface body of porous material (e.g., gravel, sand, fractured rock) filled with water and capable of supplying useful quantities of water to springs and wells.

Chemically, organic pollutants have low solubilities in water and are likely to remain as immiscible liquids (e.g., chlorinated solvents) or solids (e.g., chemical explosives) in groundwater. Numerous VOCs [e.g., *tert*-butyl ether (MTBE), *tert*-butyl alcohol (TBA)] are evident and known to accumulate in groundwater up to several 100 mg/L (Squillace and Moran 2007). Many nonreactive dissolved pollutants are transported by groundwater without the influence of any chemical, biological, or radiological processes. These are instead, influenced by the transport mechanisms listed earlier. Particulates including biologically active (e.g., bacteria, viruses), and chemically active (colloids comprising mineral and organic matter) pollutants are transported by the bulk flow of groundwater, but may be filtered out by obstructions along the flow path or by electrostatic forces on the geologic media (National Research Council 2005). Finally, *sorption* of pollutants onto solid organic matter is also of concern. Lead, for example, is a cationic metal, and commonly has diminished mobility in groundwater due to cation exchange.

1.3.3 Soils and Sediments

Assessment and monitoring of soils and sediments are more complex than for water due to such occurrences as soil erosion and sediment deposition as well as the need for extraction and digestion procedures prior to chemical analysis. Fortunately, there is an increasing trend toward the use of field-based or *in situ* techniques to help eliminate these procedures and aid in, for example, the study of element speciation and/or avoiding loss of sample integrity due to storage processes.

The analysis of soils is important for both the assessment of soil fertility and monitoring the transport of pollutants from agricultural land to rivers and coastal waters. An adequate supply of nutrients (e.g., ammonia, nitrate, phosphate) and trace elements (e.g., boron, calcium, chloride, copper, iron, sulfur, zinc) are vital for plant growth. In excess, they can be detrimental to the health of organisms and aquatic ecosystems. Ultimately, the nature and quantity of clay minerals and organic matter present in soils determine mineral solubility, pH, cation and anion exchange, buffering effects, and nutrient availability. The mobility and availability of heavy metals, for example, is strongly influenced by pH, clay content, and humus content. Redox changes are important for a number of trace elements, including arsenic (As), chromium (Cr), mercury (Hg), and selenium (Se). The full chemistry behind the redox changes is beyond the scope of this book, but it is important to note that such changes can greatly influence the fate and toxicity of these elements. Arsenic, for example, exists as As(III) and As(V), with As(III) being the more toxic of the two species. Harmful organic compounds in soils [e.g., polychlorinated biphenyls (PCBs)] are also of concern due to their long-term stability. A greater discussion of the soil environment can be found in a comprehensive work by Pierzynski et al. (2005).

Sediments are formed by the deposition of particles transported to aqueous environments, and are major sinks for nutrients, trace metals, and organic pollutants. These pollutants can be released from the sediments back into the overlying waters by the diffusion of porewater in response to changes in pH, redox potential, and salinity conditions, or as the result of bioturbation or physical resuspension (Worsfold et al. 2008).

> The term *porewater* refers to water that occupies the space between sediment particles.

It is likely that bottom-feeding organisms may take the pollutants back up and reintroduce them into the food chain. Ultimately, the quality of sediments is being increasingly recognized as a means of assessing the ecological health of aquatic ecosystems. Interim Sediment Quality Guidelines (ISQG) adapted from North American effect-based guidelines, for example, have recently been developed by Australian and New Zealand government bodies (ANZECC/ARMCANZ 2000). Selected parameters and related ISQGs are listed in Table 1.4. These values correspond to low and high percentiles of the chemical concentrations associated with adverse biological effects in field-based studies and laboratory bioassays compiled from the North American studies (Long et al. 1995).

1.4   OVERVIEW OF CHEMOMETRIC METHODS USED IN ENVIRONMENTAL ANALYSIS

Researchers involved in modern environmental analysis are faced with a multitude of tasks such as the determination of ultralow analyte levels in complex matrices, assembling baseline studies, assessing the fate and distributions of toxic chemicals, source identification, and biological response modeling. Such tasks benefit highly from an integrated approach using chemometric multivariate methods for managing

TABLE 1.4

**ANZECC/ARMCANZ Interim Sediment Quality Guidelines
for Selected Criteria/Parameters**

Sediment Quality Criteria/Parameter	ISQG—Low (Screening Level)[a]	ISQG—High (Maximum Level)[a]
Acenapthalene	16[b]	500[b]
Antimony	2	25
Arsenic	20	70
Benzo(a)pyrene	430[b]	1600[b]
Cadmium	1.5	10
Chromium	80	370
Copper	65	270
Dieldrin	0.02	8
Fluorene	19[b]	540[b]
Lead	50	220
Mercury	0.15	1
Nickel	21	52
Silver	1	3.7
Total PAHs[c]	4000[b]	45,000[b]
Tributyltin	5[d]	70[d]
Zinc	200	410

[a] All values in mg/kg dry weight unless otherwise stated.
[b] Values in µg/kg dry weight.
[c] PAHs, polycyclic aromatic hydrocarbons.
[d] Values in µg tin (Sn)/kg dry weight.

and analyzing these types of complex requests. Experimental design techniques, for example, are used to strengthen the scientific base underlying the design and application of sensitive and selective analytical instrumentation, and elucidate the biogeochemical cycles of environmentally relevant species. Pattern recognition methods are employed to reveal and evaluate complex relationships among a wide assortment of environmental applications. Multivariate classification and calibration methods are appropriate to extract predictive information from datasets that may only show small individual correlations to the property of interest. These and a variety of other innovative chemometric techniques are presented in this book through the use of theoretical and application-based approaches to environmental analysis. An overview of selected techniques in representative application areas is presented in Table 1.5, with detailed discussions in subsequent chapters.

1.5   CHAPTER SUMMARY

The ability to effectively characterize and quantify environmental pollutants of interest is of paramount importance, given the modern environmental-monitoring problems associated with the acquisition of large amounts of data. Investigators are dealing

TABLE 1.5

Selected Chemometric Applications in Environmental Analysis

Application Area	Chemometric Technique(s)	Application Notes	References
Experimental design and optimization	Factorial design, Doehlert matrix design	Utilization of factorial designs and Doehlert matrix in the optimization of experimental variables for the determination of vanadium and copper in seawater.	Ferreira et al. (2002)
Experimental design and optimization	Factorial design	Factorial designs applied to the development of a capillary electrophoresis method for the analysis of zinc, sodium, calcium, and magnesium in water samples.	Jurado-González et al. (2003)
Experimental design and optimization	Half-fractional factorial design, central composite design	Half-fractional factorial design and response surface methodology in estimating the influence of experimental variables for biomonitoring benzene in exposed individuals.	Prado et al. (2004)
Experimental design and optimization	Factorial design	Electrochemical treatment of methyl parathion based on the implementation of a factorial design.	Vlyssides et al. (2004)
Experimental design and optimization	Central composite design	Optimization of flow-injection-hydride and inductively coupled plasma spectrometric determination of selenium.	Etxebarria et al. (2005)
Experimental design and optimization	Factorial design	Factorial experimental design methodology used to study the heavy metal removal efficiency.	Hsien et al. (2006)
Experimental design and optimization	Box–Behnken design	Optimization of liquid chromatography experimental conditions for the determination of benzo[a]pyrene-quinone isomers.	Gonzalez et al. (2007)
Multivariate analysis	Cluster analysis, correlation analysis, principal component analysis	Chemometrical interpretation of lake waters after chemical analysis.	Tokalioğlu and Kartal (2002)
Multivariate analysis	Partial least-squares regression	A partial least-squares regression model was used to predict element contents in river sediments from measured contents in the particulate suspended matter.	Aulinger et al. (2004)
Multivariate analysis	Cluster analysis, multidimensional scaling, principal component analysis	Use of multivariate techniques in assessing the influence of oxygen levels and chemical pollutants on macrofaunal assemblages.	Guerra-García and García-Gómez (2005)

continued

TABLE 1.5   (continued)

Application Area	Chemometric Technique(s)	Application Notes	References
Multivariate analysis	Cluster and discriminant analysis, multilinear regression, principal component analysis	Water quality assessment and apportionment of pollution sources using multivariate statistical techniques.	Singh et al. (2005)
Multivariate analysis	Cluster analysis, factor analysis	Interpretation of river-water monitoring data using factor and cluster analysis.	Boyacioglu and Boyacioglu (2007)
Multivariate analysis	Principal component regression and artificial neural networks	Multiple regression combined with principal component analysis and artificial neural network modeling to predict ozone concentration levels in the lower atmosphere.	Al-Alawi et al. (2008)
Signal processing and time series analysis	Multivariate auto- and cross-correlation analysis	Analysis of multidimensional time series in environmental analysis.	Geiß and Einax (1996)
Signal processing and time series analysis	Simple moving average	Long-term trends of the water table fluctuations of a river basin through a time series approach.	Reghunath et al. (2005)
Signal processing and time series analysis	Data-based mechanistic models	Time-domain methods for the identification and estimation of a model for the transportation and dispersion of a pollutant in a river.	Young and Garnier (2006)
Signal processing and time series analysis	Autocorrelation function, autocovariance function	Use of signal processing in the identification and characterization of organic compounds in complex mixtures.	Pietrogrande et al. (2007)
Signal processing and time series analysis	Fast wavelet transform-based feature selection algorithm, coupled with multilinear regression- and partial least squares regression methods	New chemometric methodology to resolve, in the wavelet domain, quaternary mixtures of chlorophenols: 4-chloro-3-methylphenol, 4-chlorophenol, 2,4-dichlorophenol, and 2,4,6-trichlorophenol.	Palacios-Santander et al. (2007)
Signal processing and time series analysis	Recurrence quantification analysis based on recurrence plots	A nonlinear time series analysis technique for the characterization of regime shifts in environmental time series.	Zaldivar et al. (2008)

with ever-increasing levels of legislation and reporting, but are spurred on by public interest and a concerned global community. The demand to characterize and analyze complex environmental matrices and determine concentrations at ultralow levels has inspired the need for more sensitive and selective analytical methods. Modern chemometric methods are becoming increasingly used in such endeavors, providing a strong tool to perform well-planned and carefully designed experiments, recognizing patterns and structures in environmental datasets, and helping to evaluate the relationships that exist among a wide range of environmental applications.

1.6   END OF CHAPTER PROBLEMS

1. In your own words, define chemometrics and describe how such techniques can be used to aid in modern environmental analysis.
2. List three steps that humans can take to help reduce ground-level ozone formation.
3. Provide a detailed chemical mechanism by which nitrogen dioxide (NO_2) is converted to nitric acid (HNO_3) in the atmosphere.
4. Describe the differences between point and nonpoint sources of pollution and provide examples of each.
5. Why is monitoring of "total" metals in aqueous environments not accepted as the best practice?
6. How does the use of nitrate and phosphate as fertilizers lead to problems with rivers, streams, and coastal waters?
7. How do acidified lakes and streams act to increase the risks to the environment posed by metals such as cadmium, lead, and mercury?
8. How does acid precipitation affect the solubility of metals in soils?
9. Explain how sediments can be both a carrier and a source of pollutants in aquatic systems.

REFERENCES

Al-Alawi, S. M., Abdul-Wahab, S. A., and Bakheit, C. S. 2008. Combining principal component regression and artificial neural networks for more accurate predictions of ground-level ozone. *Environmental Modelling & Software* 23: 396–403.

American Lung Association. 2006. Website. Accessed November 12, 2007.

Apte, S. C., Batley, G., and Maher, W. A. 2002. Monitoring of trace metals and metalloids in natural waters. In *Environmental Monitoring Handbook*, ed. F. R. Burden, I. D. McKelvie, U. Forstner and A. Guenther, 6.1–6.27. New York: McGraw-Hill.

Aulinger, A., Einax, J. W., and Prange, A. 2004. Setup and optimization of a PLS regression model for predicting element contents in river sediments. *Chemometrics and Intelligent Laboratory Systems* 72: 35–41.

Australian and New Zealand Environment Conservation Council/Agriculture and Resource Management Council of Australia and New Zealand (ANZECC/ARMCANZ). 2000. *Australian and New Zealand Guidelines for Fresh and Marine Water Quality*. Canberra, NSW.

Blount, B. C., Pirkle, J. L., Osterloh, J. D., Valentin-Blasini, L., and Caldwell, K. L. 2006. Urinary perchlorate and thyroid hormone levels in adolescent and adult men and women living in the United States. *Environmental Health Perspectives* 114: 1865–1871.

Boulding, J. R. 1995. *Practical Handbook of Soil, Vadose Zone, and Ground-water Contamination: Assessment, Prevention, and Remediation*. Boca Raton: Lewis Publishers.

Boyacioglu, H. and Boyacioglu, H. 2007. Surface water quality assessment by environmental methods. *Environmental Monitoring and Assessment* 131: 371–376.

Brereton, R. G. 1990. *Chemometrics: Applications of Mathematics and Statistics to Laboratory Systems*. New York: Ellis Horwood.

Brown, S. D., Sum, S. T., Despagne, F., and Lavine, B. K. 1996. Chemometrics. *Analytical Chemistry* 68: 21–61.

Butcher, S. S., Charlson, R. J., Orians, G. H. and Wolfe, G. V. 1992. *Global Biogeochemical Cycles*. San Diego: Academic Press.

Cheremisinoff, N. P. 2002. *Handbook of Air Pollution Prevention and Control*. New York: Butterworth Heinemann.

Commission of the European Communities. 2005. Communications from the Commission to the Council and the European Parliament – Thematic Strategy on Air Pollution. Brussels: COM(2005) Final.

Currie, R. A., Filliben, J. J., and DeVoe, J. R. 1972. Statistical and mathematical methods in analytical chemistry. *Analytical Chemistry* 44: 497–512.

Donat, J. R. and Bruland, K. W. 1995. Trace elements in the oceans. In *Trace Elements in Natural Waters*, ed. B. Salbu and E. Steines, 247–292. Boca Raton: CRC Press.

Etxebarria, N., Antolín, R., Borge, G., Posada, T., and Raposo, J. C. 2005. Optimization of flow-injection-hydride generation inductively coupled plasma spectrometric determination of selenium in electrolytic manganese. *Talanta* 65: 1209–1214.

Ferreira, S. L. C., Queiroz, A. S., Fernandes, M. S., and dos Santos, H. C. 2002. Application of factorial designs and Doehlert matrix in optimization of experimental variables associated with the preconcentration and determination of vanadium and copper in seawater by inductively coupled plasma optical emission spectrometry. *Spectrochimica Acta* 57: 1939–1950.

Geiβ, S. and Einax, J. 1996. Multivariate correlation analysis—a method for the analysis of multidimensional time series in environmental studies. *Chemometrics and Intelligent Laboratory Systems* 32: 57–65.

Gonzalez, A., Foster, K. L., and Hanrahan, G. 2007. Method development and validation for optimized separation of benzo[a]pyrene-quinone isomers using liquid chromatography-mass spectrometry and chemometric response surface methodology. *Journal of Chromatography A* 1167: 135–142.

Guerra-García, J. M. and García-Gómez, J. C. 2005. Oxygen levels versus chemical pollutants: do they have similar influence on macrofaunal assemblages? A case study in a harbor with two opposing entrances. *Environmental Pollution* 135: 281–291.

Hanrahan, G., Gledhill, M., House, W. A., and Worsfold, P. J. 2001. Phosphorus loading in the Frome Catchment, UK: seasonal refinement of the coefficient modelling approach. *Journal of Environmental Quality* 30: 1738–1746.

Hanrahan, G., Gardolinski, P., Gledhill, M., and Worsfold, P. J. 2002. Environmental monitoring of nutrients. In *Environmental Monitoring Handbook*, ed. F. R. Burden, I. D. McKelvie, U. Forstner, and A. Guenther, 8.1–8.16. New York: McGraw-Hill.

Hanrahan, G., Salmassi, T. M., Khachikian, C. S., and Foster, K. L. 2005. Reduced inorganic phosphorus in the natural environment: significance, speciation and determination. *Talanta* 66: 435–444.

Hart, B. T. 2002. Water quality guidelines. In *Environmental Monitoring Handbook*, ed. F. R. Burden, I. D. McKelvie, U. Forstner, and A. Guenther, 1.1–1.26. New York: McGraw-Hill.

Hill, M. K. 2004. *Understanding Environmental Pollution*. Cambridge: Cambridge University Press.

Hopke, P. K. 2003. The evolution of chemometrics. *Analytica Chimica Acta* 500: 365–377.

Hsien, I., Kuan, Y.-C., and Chern, J.-M. 2006. Factorial experimental design for recovering heavy metals from sludge with ion-exchange resin. *Journal of Hazardous Materials* 138: 549–559.

Jurado-González, J. A., Galindo-Riaño, M. D., and García-Vargas, M. 2003. Factorial designs applied to the development of a capillary electrophoresis method for the analysis of zinc, sodium, calcium and magnesium in water samples. *Talanta* 59: 775–783.

Kongtip, P., Thongsuk, W., Yoosook, W., and Chantanakul, S. 2006. Health effects of metropolitan traffic-related air pollutants on street vendors. *Atmospheric Environment* 40: 7138–7145.

Kowalski, B. R. 1975. Chemometrics: views and propositions. *Journal of Chemical Information and Computer Sciences* 15: 201–203.

Lavine, B. and Workman, J. J. 2004. Chemometrics. *Analytical Chemistry* 76: 3365–3372.

Long, E. R., MacDonald, D. D., Smith, S. L., and Calder, F. D. 1995. Incidence of adverse biological effects within ranges of chemical concentrations in marine and estuarine sediments. *Environmental Management* 19: 81–97.

Massart, D. L., Vandeginste, B. G. M., Buydens, L. M. C., De Jong, S., Lewi, P. J., and Smeyers-Verbeke, J. 1997. *Handbook of Chemometrics and Qualimetrics, Part A.* Amsterdam: Elsevier Science.

National Research Council. 2005. *Contaminants in the Subsurface.* Washington D.C.: The National Academy Press.

Novello, D. P. and Martineau, R. J. 2005. *The Clean Air Act Handbook*, 2nd Edition, New York: American Bar Association.

Palacíos-Santander, J. M., Cubillana-Aguilera, L. M., Naranjo-Rodríguez, I. and Hidalgo-Hidalgo-Cisneros, J. L. 2007. A chemometric strategy based on peak parameters to resolve overlapped electrochemical signals. *Chemometrics and Intelligent Laboratory Systems* 85: 131–139.

Pierzynski, G. M., Sims, J. T., and Vance, G. F. 2005. *Soils and Environmental Quality*, 3rd Edition, Boca Raton: Taylor & Francis.

Pietrogrande, M. C., Mercuriali, M., and Pasti, L. 2007. Signal processing of GC–MS data of complex environmental samples: characterization of homologous series. *Analytica Chimica Acta* 594: 128–138.

Prado, C., Garrido, J., and Periago, J. F. 2004. Urinary benzene determination by SPME/GC-MS. A study of variables by fractional factorial design and response surface methodology. *Journal of Chromatography B* 804: 255–261.

Pyne, S. 2002. Small particles add up to big disease risk. *Science* 295: 1994–1997.

Reghunath, R., Murthy, T. R. S., and Raghavan, B. R. 2005. Time series analysis to monitor and assess water resources: a moving average approach. *Environmental Monitoring and Assessment* 109: 65–72.

Reynolds, C. S. 1984. *The Ecology of Freshwater Phytoplankton.* Cambridge: Cambridge University Press.

Singh, K. P., Malik, A., and Sinha, S. 2005. Water quality assessment and apportionment of pollution sources of Gomti River (India) using multivariate statistical techniques— a case study. *Analytica Chimica Acta* 538: 355–374.

Squillace, P. J. and Moran, M. J. 2007. Factors associated with sources, transport, and fate of volatile organic compounds and their mixtures in aquifers of the United States. *Environmental Science and Technology* 41: 2123–2130.

Sullivan, P. J., Agardy, F. J., and Clark, J. J. 2005. *The Environmental Science of Drinking Water.* Amsterdam: Elsevier.

Tokalioğlu, S. and Kartal, S. 2002. Chemometrical interpretation of lake waters after their chemical analysis by using AAS, flame photometry and titrimetric techniques. *International Journal of Environmental Analytical Chemistry* 82: 291–305.

U.S. EPA. 1994. *Water Quality Standards Handbook*, 2nd Edition, EPA 823-B-94-005a, Washington, D.C.

Van Nieuwenhuyse, E. E. and Jones, J. R. 1996. Phosphorus-chlorophyll relationship in temperate streams and its variation with stream catchment area. *Canadian Journal of Fisheries and Aquatic Sciences* 53: 99–105.

Vlyssides, A. G., Arapoglou, D. G., Israilides, C. J., Barampouti, E. M. P., and Mai, S. T. 2004. Electrochemical treatment of methyl parathion based on the implementation of a factorial design. *Journal of Applied Electrochemistry* 34: 1265–1269.

Wold, S. and Sjöström, M. 1998. Chemometrics, present and future success. *Chemometrics and Intelligent Laboratory Systems* 44: 3–14.

WHO. 2006. *Guidelines for drinking water quality*, 3rd Edition, Geneva.

Worsfold, P. J., McKelvie, I. D., and Hanrahan, G. 2008. Environmental applications of flow injection analysis—waters, sediments and soils. In *Advances in Flow Injection Analysis and Related Techniques*, ed. S. D. Kolev and I. D. McKelvie. Chichester: Elsevier.

Young, P. C. and Garnier, H. 2006. Identification and estimation of continuous-time, data-based mechanistic (DBM) models for environmental systems. *Environmental Modelling & Software* 21: 1055–1072.

Zaldivar, J.-M., Strozzi, F., Dueri, S., Marinov, D., and Zbilut, J. P. 2008. Characterization of regime shifts in environmental time series with recurrence quantification analysis. *Ecological Modelling* 210: 58–70.

Zhang, H. and Davison, W. 2001. In situ speciation measurements using diffusive gradients in thin films (DGT) to determine inorganically and organically complexed metals. *Pure and Applied Chemistry* 73: 9–15.

2 Review of Statistics and Analytical Figures of Merit

2.1 ERROR AND SAMPLING CONSIDERATIONS

Measurements play a dominant role in quantitative analysis and are subject to experimental error. *Uncertainty* in the results can be minimized, but never completely eliminated. This trait is inherent in any field where numerical results are obtained. It is especially important in environmental analysis where large datasets from repeated measurements of a sample or group of samples are obtained to understand error sources and treat uncertainty. Before we delve into the types of error encountered in environmental analysis, let us first step back and consider the *sample* in both a statistical sense and an environmental sense.

> *Uncertainty* can be defined as "the interval around the result of the measurement that contains the true value with high probability" (Ramsey 2002).

Statistically, a sample can be defined as a group of objects selected from the *population* of all such objects (Miller and Miller 2005). When sampling from a population, the sample should, ideally, be *representative* for a particular attribute. This is often achieved by taking numerous *random* samples (within the time and financial constraints of the project). In environmental analysis, you will also encounter the word *sample* as referring to the actual material being studied (e.g., a water sample), which denotes one sample of water (e.g., a 100 mL container) from a larger body of water. By studying this sample, it is hoped to draw valid conclusions about, for example, the physicochemical characteristics of the larger body of water.

> *Population*, in an environmental context, refers to the entire collection of samples of air, soil, water, organisms, or other material that could potentially be sampled.

We commonly think of the *random errors* and *systematic errors* associated with measurements after performing *replicate* analyses, but often neglect those that can occur during the sampling/sample preparation process. Consider a sample of measurements of total cadmium concentration from an infinite population of all such possible measurements in a heterogeneous mixture. Let us first ponder sample size—an important factor when considering sampling precision. Work by Ingamells and Switzer (1973) provided evidence for preliminary analysis of at least three samples when studying heterogeneous materials. They showed that a relationship between sample size (w) and the percent standard deviation (R) exists, such that

$$wR^2 = K \tag{2.1}$$

When K is known, the value of w required to achieve a desired precision can be computed. K is obtained by the investigator working on three or more sets of samples, each set at a different level of w. If the variance R^2 is plotted against $1/w$, K can be obtained from the slope. Note the necessity for the sample size to be large enough for a reasonable probability to exist that each type of attribute in the population is present in sample increments.

> *Replicate* refers to samples of similar proportion and composition carried through an analysis in exactly the same way in a repeated fashion.

Pierre Gy has provided key evidence as to why investigators need to be critically aware of the sampling process they design, and eloquently described the errors associated in the complete environmental analysis process (Gy 2004a,b). To accurately assess the uncertainty of environmental analyses, we must examine the global estimation error (GEE), which comprises both the total sampling error (TSE) and the total analytical error (TAE). Here, the various forms of heterogeneity in the sample must be defined before effectively determining the mean, variance, and mean square. Therefore, the results used to estimate sampling uncertainty must be obtained from carefully designed experiments for interpretation by ANOVA (*analysis of variance*) methods (Thompson 1998). Section 2.8 presents an expanded discussion of ANOVA, and Chapter 4 describes experimental design procedures. Detailed calculations of the errors discussed here are provided in Gy (2004a). Generalized errors in environmental analysis are highlighted in Table 2.1, with sampling strategies described in Chapter 3.

2.2 DESCRIPTIVE STATISTICS

In environmental analysis, we are interested in the characteristics of finite populations. Consider the following example and the terms presented. Daily readings of mean nitrate levels are taken in the form of a random sample (with subsequent laboratory analysis) of, let us assume, 5 of the 20 monitoring stations around a lake

TABLE 2.1

Types of Errors, Examples, and Characteristics

Type of Error	Examples	Causes/Characteristics
Random	An investigator measures the mass of a soil sample three times using the same balance with differing values: 10.34 g, 10.44 g, 10.37 g.	Arises from limitations in an investigator's ability to make physical measurements, or from natural fluctuations in the quantity being measured.
	Random electrical noise in an instrumental response.	Can be reduced but not eliminated.
		Individual results fall on both sides of the average value.
		Can be estimated using replicate measurements.
Systematic	An investigator uses two different sampling procedures when estimating mercury concentrations in a contaminated lake.	Arises from both human and equipment/instrument causes.
		Cannot be detected simply by performing replicate measurements.
	Using a pH meter that has not been calibrated correctly.	Causes all results to be either lower or higher than the average.

(see Excel® Spreadsheet exercise in Section 2.13). The sample mean level gives an estimate of the population mean nitrate level for the lake as a whole, but is met with a high degree of sampling uncertainty. Here, the sample mean is the average value of a sample, which is a finite series of measurements and is calculated as

$$\overline{x} = \frac{\sum x_i}{n} \qquad (2.2)$$

where n = sample size and x_i = individual measured values. To measure the spread of the individual values, we calculate the sample *standard deviation* (s) as

$$s = \sqrt{\frac{\sum_i (x_i - \overline{x})^2}{n-1}} \qquad (2.3)$$

where the quantity $n - 1$ = the degrees of freedom (DOF). This term often confuses novice investigators, and refers to the number of mutually independent deviations, represented by $(x - \overline{x})$, used in calculating the standard deviation. For an infinite set of data, the population standard deviation (often called the true standard deviation) is represented by σ. The International Union of Pure and Applied Chemistry (IUPAC) has established the rules by which s is used when the number of replicates is smaller than 10 (IUPAC 1997). Generally, s becomes a more reliable expression of precision (see discussion on Ingamells and Switzer, Section 2.1) as n becomes larger. The relative standard deviation (RSD) is much more convenient and widely reported.

It is expressed as a percentage (%) and is obtained by multiplying the sample standard deviation by 100 and dividing this product by the sample mean

$$RSD = \frac{100\,s}{\bar{x}}$$

(2.4)

This method of reporting is a way to express the random error present regardless of analyte concentration or weight measured (Haswell 1992). Additionally, s can be expressed as variance (the square of the standard deviation). The *median* is the middle number in a given series of measurements. If the measurements are of an odd number, the index of the median is represented by $\frac{1}{2}(n + 1)$. For an even number, the median is represented by $\left(\frac{1}{2}n + 1\right)$. Note that this assumes that the measurements are ordered (increasing or decreasing). The *range* is the difference between the highest and lowest values.

2.3   DISTRIBUTION OF REPEATED MEASUREMENTS

We learned from the previous section that standard deviation gives a measure of the spread of a particular set of data about the mean value. Unfortunately, standard deviation does not give us the shape of the distribution of the data. Consider an investigator who has performed 50 replicate measurements of phosphate ion concentration (μM) in a river-water sample. The results of all 50 measurements are presented in Table 2.2.

Spread is the difference between the highest and lowest results of a set of measurements.

EXAMPLE PROBLEM 2.1

1. Suppose the measured nitrate concentrations (μM) from the five stations of the lake were as follows: 3.4, 3.9, 4.1, 3.6, and 3.7. Calculate the sample mean and standard deviation.

Answer: The mean nitrate concentration (μM) is

$$\bar{x} = \frac{3.4 + 3.9 + 4.1 + 3.6 + 3.7}{5} = 3.74$$

$$s = \sqrt{\frac{(3.4 - 3.74)^2 + (3.9 - 3.74)^2 + (4.1 - 3.74)^2 + (3.6 - 3.74)^2 + (3.7 - 3.74)^2}{5 - 1}} = 0.27$$

The results are then reported as follows: 3.74 μM $\pm$ 0.27 μM.

As can be seen in Table 2.2, all 50 measurements are visible, but the frequency at which each appears is not apparent. If the investigator constructs a *frequency table*

TABLE 2.2
Results from 50 Determinations of Phosphate Ion Concentration (μM) in a River-Water Sample

0.73	0.74	0.73	0.74	0.73	0.73	0.76	0.76	0.74	0.70
0.72	0.76	0.77	0.72	0.73	0.73	0.72	0.69	0.73	0.72
0.73	0.71	0.73	0.71	0.73	0.78	0.71	0.74	0.71	0.72
0.75	0.71	0.75	0.75	0.74	0.72	0.73	0.72	0.74	0.74
0.75	0.74	0.74	0.70	0.73	0.72	0.75	0.74	0.75	0.75

(Table 2.3), the frequency of each measured concentration in the sample can be obtained. For further clarification, the investigator can construct a *histogram*, which is a nonparametric estimate of a probability distribution (see the following discussion on nonparametric methods), of the phosphate ion concentration data in Table 2.2. The histogram (Figure 2.1a) shows with a few exceptions, that the distribution of the measurements is roughly symmetrical about the mean value (0.733 μM). The normal quantile plot (Figure 2.1b) adds a graph to the results for visualizing the extent to which phosphate ion measurements are normally distributed. Quantiles are values that divide a distribution into two groups where the pth quantile is larger than $p\%$ of the values. As shown, the quantile plot approximates a diagonal straight line, a sign of normal distribution. The x-axis shows the expected normal scores for each value. Also shown are Lilliefors confidence bounds, reference lines, and a probability scale (top axis).

The quantile box plot (Figure 2.2a) can often show additional quantiles (90th, 95th, or 99th) on the response axis. If a distribution is normal, the quantiles shown in a box plot are approximately equidistant from each other. This allows one to see

TABLE 2.3
Frequency Table for the 50 Phosphate Ion Concentration (μM) Measurements in a River-Water Sample

Phosphate Ion Concentration (μM)	Frequency
0.69	1
0.70	2
0.71	5
0.72	8
0.73	12
0.74	10
0.75	7
0.76	3
0.77	1
0.78	1
	Total = 50

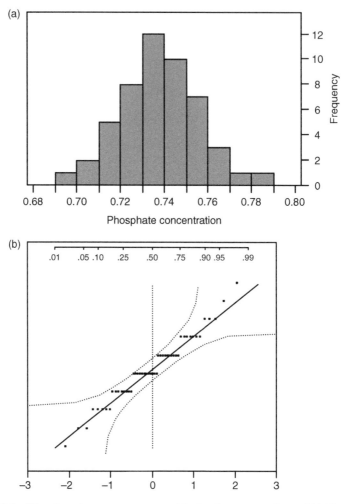

FIGURE 2.1   Phosphate ion measurements with visual representations of (a) histogram and (b) normal quantile plot.

at a glance whether the distribution is normal or not. The difference between the scores at the 75th and 25th percentiles is called the *H*-spread. The vertical lines that extend above and below the 75th and 25th percentiles are called whiskers and the horizontals at the end of the whiskers are termed adjacent values. The quantile table (Table 2.4) lists the quantile distances for phosphate ion measurements. An outlier box plot can be overlaid on the quantile box plot (Figure 2.2b) for detection (visual) of potential outlying data points. Generally, any points that lie beyond the adjacent values are graphed using small circles (0.78 on Figure 2.2b). More detailed information on outlier detection can be found in Section 2.7.

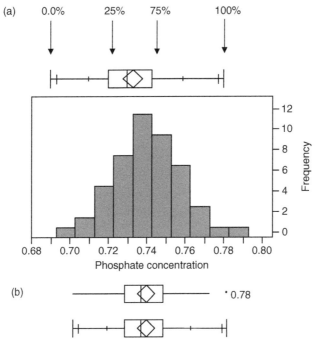

FIGURE 2.2 Phosphate ion measurement distribution showing (a) the quantile box plot and (b) overlay of the outlier box plot and the quantile box plot.

TABLE 2.4
Quantile Spacing of Phosphate Ion Data

Quantiles (%)	Designation	Phosphate Ion Concentration (μM)
100	Maximum	0.78
99.5		0.78
97.5		0.77
90.0		0.76
75.0	Quartile	0.74
50.0	Median	0.73
25.0	Quartile	0.72
10.0		0.71
2.5		0.69
0.5		0.69
0.0	Minimum	0.69

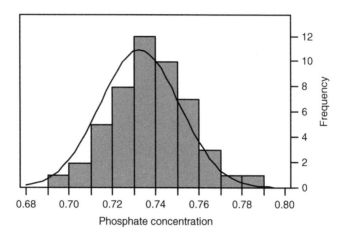

FIGURE 2.3 Plot of the mathematical function for a nearly normal distribution of phosphate ion measurements.

Being able to describe the distribution of measurements using a mathematical formula has obvious advantages in describing the form of the population. Here, a continuous random variable (x) has a normal (Gaussian) distribution around the mean of the population. For a normal distribution, ~68% of the population lies within $\pm1\sigma$ of the mean, ~95% within $\pm2\sigma$, and ~99.7% within $\pm3\sigma$. Consider again the investigator's results for phosphate ion determination (Figure 2.3). The smooth curve superimposed on the histogram is a plot of the mathematical function for a nearly ideal normal distribution with a population mean μ of 0.733 μM and a population standard deviation σ of ±0.018. If we assume that the phosphate concentrations are normally distributed, then 68% would lie in the range 0.683–0.783 μM, 95% within 0.663–0.803 μM, and 99.7% between 0.660 and 0.806 μM.

A cumulative distribution function (CDF) plot (Figure 2.4) can be generated using the observed data as it relates to the density curve in Figure 2.3. This plot completely describes the probability distribution and estimates the areas under the density curve up to the point x_i.

If investigators need to test whether the overall shape of the frequency distribution differs significantly from the normal, goodness-of-fit tests can be employed. The chi-square test is an often employed technique that is based on

$$X^2 = \sum_i \frac{(O_i - E_i)^2}{E_i} \tag{2.5}$$

where O_i = observed frequencies and E_i = expected frequencies according to some *null hypothesis* (e.g., the data follows a specified distribution). See Section 2.6 for a detailed discussion on the null hypothesis. This particular test requires a sufficient sample size in order for the chi-square approximation to be valid.

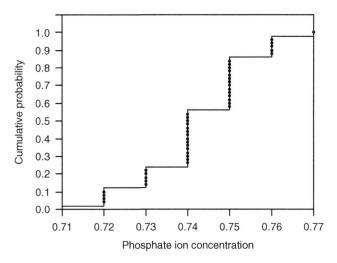

FIGURE 2.4   CDF plot of phosphate-ion measurements.

Another tool is the stem and leaf plot (Figure 2.5), an alternative to the histogram. Each line of the plot has a stem value that is the leading digit of a range of column values. The leaf values are made from the next-in-line digits of the values. One can reconstruct the data values by joining the stem and leaf with the aid of the legend at the bottom of the plot.

What if a particular set of data is not normally distributed? One can again refer to the sample size to help answer this question. However, it may not be possible to perform 50 replicates as the investigator has done here. Developed by statisticians, the *central limit theorem* states that as the number of random samples increases (≥30),

Stem	Leaf	Count
78	0	1
77	0	1
76	000	3
75	0000000	7
74	0000000000	10
73	000000000000	12
72	00000000	8
71	00000	5
70	00	2
69	0	1

69|0 represents 0.690

FIGURE 2.5   Stem and leaf plot of phosphate ion measurements.

the sampling distribution of the mean will likely take on a normal distribution, even if the original population is not normal. Essentially, as sample size increases, the sampling distribution of the sample means resembles a normal distribution, with a mean the same as that of the population and a standard deviation equal to the standard deviation of the population divided by the square root of n. For smaller replicate measurements (≤ 30), investigators use $n - 1$ (remember the discussion on DOF). It is recommended that a minimum of six replicates, corresponding to five values ($6 - 1 = 5$), be performed to obtain a reliable standard deviation (Haswell 1992). See Section 2.12 for a research-based application of sample size, and its effects on relative errors in normal and log-normal distributions. For the latter, frequency is plotted against the logarithm (e.g., to the base 10) of the characteristic (e.g., concentration) and gives an approximate normal distribution curve. Such a method is often applied to particle-size distribution studies. For example, atmospheric aerosol distributions are often complex due to their large numbers and varied sources. Statistical fitting with two or three log-normal modes, for example, can aid overall classification.

From the preceding discussions, the need is evident for statistical procedures that allow investigators to process data on variables about which little or nothing is known (e.g., regarding their distribution). Specifically, nonparametric methods are utilized in cases where the researcher knows nothing about the parameters of the variable of interest in the population. Note that parametric statistical methods refer to those that require a parametric assumption such as normality. Nonparametric methods do not rely on the estimation of parameters (e.g., mean or standard deviation) describing the distribution of the variable of interest in the population. Therefore, these are also sometimes called parameter-free or distribution-free methods. Fortunately, there is at least one nonparametric equivalent for each parametric type of test and they fall into the following categories:

1. Tests of differences among groups—Wald–Wolfowitz runs test, the Mann–Whitney U test, and the Kolmogorov–Smirnov two-sample test;
2. Tests of differences among variables—Sign test, Wilcoxon's matched pairs test, and Cochran Q test;
3. Tests of relationships among variables—Spearman's Rho, Kendall's Tau, and Hoeffding's D tests.

2.4   DATA QUALITY INDICATORS

Data quality indicators (DQIs), as characterized by the United States Environmental Protection Agency (U.S. EPA), refer to statements of data quality commonly used to express measurement uncertainty as: accuracy, precision, representativeness, completeness, and comparability (U.S. EPA 1997). These parameters can be used to evaluate both the qualitative and quantitative components of total error discussed in Section 2.1, and to help establish quality assurance procedures (Chapter 3). They are used to determine if the data quality objectives (DQOs) established early on in the study have been met. The DQO process defines the problem to be studied, and the type, quality, and quantity of the data to be gathered and processed.

Secondary DQIs (those not always utilized in data quality evaluation) also have an effect on the results of qualitative and quantitative analysis and include: repeatability, reproducibility, recovery, sensitivity, and limit of detection. The last two are discussed in Section 2.10. Ultimately, secondary DQIs can affect the outcome of primary DQIs and thus, should not be discussed as separate items. Information on verification, validation, and integrity—which, although not classified as indicators, directly influence data quality—is discussed in Chapter 3.

2.4.1 PRIMARY DQIs

Both *precision* and *accuracy* comprise error. Precision (random error) describes the reproducibility of measurements and is generally determined by repeating the measurements on replicate samples. It can be calculated as the standard deviation or relative standard deviation using Equations 2.3 and 2.4. Additionally, we can use the relative percent difference (RPD) to measure the precision between two duplicate measurements such that

$$\%RPD = \frac{x_i - x_d}{\bar{x}} \times 100 \tag{2.6}$$

where x_i is the sample measurement, x_d the duplicate measurement, and $\bar{x}$ is the mean of the sample and duplicate. It is difficult to assess the level of precision as this depends on the sample type and methodology adopted. It is generally accepted that RSD values of $<5.0\%$ are acceptable, with values above that indicating the presence of random errors. In addition, when assessing these random errors, it is recommended to have a minimum of six (often upwards of 10) replicate values taken (Haswell 1992).

Accuracy measures the closeness of a measurement to an accepted reference or true value.

Precision is the agreement among individual measurements of the same property under a prescribed set of conditions.

Bias is systematic distortion of a measurement process that causes error in one direction.

Accuracy is the combination of random error (precision) and systematic error (bias). It can be measured by determining the percent recovery of an analyte that has been added (spiked) to samples before analysis at a known concentration. Percent recovery (R) is calculated as

$$\%R = \frac{x_{s+i} - x_i}{x_t} \times 100 \tag{2.7}$$

where x_{s+i} is the spiked sample measurement, x_i the sample measurement, and x_t is the actual concentration of spike added.

EXAMPLE PROBLEM 2.2

1. Is it possible to have high precision in your measurements while at the same time exhibiting poor accuracy?

Answer: Yes. However, there must be a consistent bias (e.g., the lack of representativeness) present. An illustration of this answer is represented by the targets below. The left target shows how we can achieve high precision (three replicate darts are close to each other) with poor accuracy (darts far from the desired bullseye of the target). The target on the right illustrates both high accuracy and precision (a much better score!).

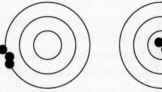

As mentioned in Example Problem 2.2, lack of *representativeness* can cause bias in measurements. Representativeness is perhaps one of the most relevant indicators in environmental analysis as it encompasses accuracy, precision, and completeness. It is a qualitative term depending heavily on the sample planning stage. Even the best sampling designs (Chapter 3), if conceived and implemented poorly, can produce unrepresentative data. *Comparability* is also qualitative in nature and expresses the similarity of attributes in a given dataset. Data comparability is possible when data are of known quality and can thus be validly applied by external users. Comparability can be enhanced by the use of intercomparison studies (Chapter 3) and the use of the same reporting units and method conditions.

Representativeness is a measure of the degree to which data accurately and precisely represent a population parameter as a sampling point, or a process condition.

Completeness is the percentage of valid data points relative to the total population of data points.

Comparability is the measure of confidence that two or more data sets may contribute to a common analysis.

(U.S. EPA 1997)

2.4.2 SECONDARY DQIs

The concepts of *repeatability* and *reproducibility* are similar in that they give us an idea of the closeness of agreement among individual results obtained with the same method on identical test material. Where they differ is in the conditions of analysis. Repeatability gives us the closeness of agreement under the same conditions (e.g., operator, time interval, test laboratory), while reproducibility shows the closeness of agreement under different sets of conditions. Both are normally expressed as the variance or standard deviation. A low variance or standard deviation, for example, indicates a high degree of reproducibility.

Recovery (*R*) is a vital indicator in environmental analytical methods as it has the capability to measure bias in a measurement process. It is used to indicate the yield of an analyte in the preconcentration or extraction stage of an analytical process. The recovery is computed as (Burns et al. 2002)

$$R_A = \frac{Q_A \text{ (yield)}}{Q_A \text{ (original)}} \tag{2.8}$$

where Q_A (original) is the known original and Q_A (yield) the recovered quantity of analyte A. Apparent recovery is advisable when calculating via a calibration graph (Section 2.9).

2.5 CONFIDENCE INTERVALS

We know from our discussions in Sections 2.2 and 2.3 that it is impossible to find the true mean, μ, and true standard deviation, σ, from a limited number of measurements. What we can do is use a sample to define a range (*confidence interval*) in which there is a specified probability of finding μ within a certain distance from the measured mean. The extreme values of this range are called *confidence limits*. Assuming that the sample size is ≤30, we have (Harris 2005)

$$\mu = \bar{x} \pm \frac{ts}{\sqrt{n}} \tag{2.9}$$

where s is the measured standard deviation, n the number of measurements, and t is the Student's t, obtained from Table 2.5. The Student's t is a statistical tool used to express confidence intervals. Note that in this table, the DOF are equal to $n - 1$. If there are six data points, for example, the DOF are five. If independent repeated samples are taken from the same population with a confidence interval calculated for each, a certain percentage (confidence level) of the intervals includes the unknown population parameter. The confidence level is the probability value $(1 - \alpha)$ associated with the confidence interval. The level of α can be used to determine the extent of the confidence interval or for direct comparison with a single p-value (or probability). Statistical significance is assumed if the calculations yield a p-value that is below α, or a $1 - \alpha$ confidence interval whose range excludes the null result.

Confidence intervals are routinely calculated at 95%, but 50%, 90%, 99%, 99.5%, and 99.9% confidence intervals are also reported. The width of the confidence interval gives us some idea about how uncertain we are regarding the unknown parameter (see discussion on precision). Investigators faced with rather wide intervals may want to consider collecting additional data before making any definite report about the parameter. The confidence interval report for the phosphate ion data (Section 2.3) is

TABLE 2.5

Values of Student's *t* at the Various Confidence Levels

	Confidence Level and *p*-Values (in Parentheses)[a]					
Degrees of Freedom	90% (0.10)	95% (0.05)	98% (0.02)	99% (0.01)	99.5% (0.005)	99.9% (0.001)
1	6.31	12.71	31.82	63.66	127.3	636.6
2	2.92	4.30	6.96	9.92	14.08	31.60
3	2.35	3.18	4.54	5.84	7.45	12.92
4	2.13	2.78	3.74	4.60	5.59	8.61
5	2.02	2.57	3.36	4.03	4.77	6.87
6	1.94	2.44	3.14	3.71	4.31	5.96
7	1.89	2.36	2.99	3.50	4.02	5.41
8	1.86	2.31	2.90	3.36	3.83	5.04
9	1.83	2.26	2.82	3.25	3.69	4.78
10	1.81	2.22	2.76	3.17	3.58	4.59
25	1.70	2.06	2.48	2.78	3.08	3.45
50	1.68	2.01	2.40	2.68	2.94	3.26

[a] Critical values of $|t|$ listed are for a two-tailed test. For a one-tailed test, the value is taken from the column twice the desired *p*-value.

presented in Table 2.6. The report shows the mean and standard deviation parameter estimates with upper and lower confidence limits for $1 - \alpha$.

2.6 STATISTICAL TESTS

The determination of statistical significance is the cornerstone of any analytical method and has an obvious importance in quantitative environmental analysis. We have touched upon this concept briefly in Section 2.5, learning that significance is assumed if the calculations yield a *p*-value that is below α, or a $1 - \alpha$ confidence interval whose range excludes the null result. Here, a test of the truth of the null hypothesis (H_0) is performed. Ostensibly, we ask how likely it is that the null hypothesis is true. The answer is usually in the form of a *p*-value. The error of H_0 being rejected when it is actually true is termed a Type I error. Conversely, the error made

TABLE 2.6

Confidence Interval Report for the Phosphate Ion Data Reported in Section 2.3

Parameter	Estimate	Lower CI	Upper CI	$1 - \alpha$
Mean	0.733	0.727	0.738	0.950
Standard deviation	0.018	0.015	0.022	—

when H_0 is false (but accepted as true) is termed a Type II error. The latter is more prevalent when α is decreased.

If we assume that the null hypothesis is true, statistical analysis can be used to calculate the probability that the observed difference between the sample mean and population mean is due solely to random errors. In general, a p-value of >0.05 indicates insufficient evidence to say that they differ. A p-value of <0.05 indicates significant difference and thus ample evidence to say they do in fact differ. It is important to note that to reject a hypothesis is to conclude that it is false. Be critically aware that to accept a hypothesis does not mean that it is true, only that there is insufficient evidence to deem otherwise. Thus, hypotheses tests usually comprise both a condition that is doubted null hypothesis) and one that is believed (alternative hypothesis, H_1 or H_A).

There are two types of tests for H_1, one-tailed and two-tailed. In a one-tailed test, the area associated with α is placed in either one tail or the other ($+$ or $-$) of the sampling distribution. Consider the value of t_{crit} to be positive and set to 0.05 with nine DOF. Examination of Table 2.5 reveals a t_{crit} of $+1.83$ (keeping in mind that the value is taken from the column twice the desired p-value for a one-tailed test). Alternatively, consider a two-sided test where α is placed in both the left ($-$) and right ($+$) tails. Consider again a value of t_{crit} set to 0.05 with nine DOF Table 2.5 reveals a t_{crit} of 2.26.

EXAMPLE PROBLEM 2.3

1. An investigator analyzed a series of sediment samples ($n = 7$) from a potentially contaminated harbor and found that the mean nickel concentration was 56 mg/kg dry weight with a standard deviation of $s = 3.5$. We learned from Table 1.3 that the interim sediment quality guideline (ISQG) for nickel at the maximum level is 52 mg/kg dry weight. Is the sediment contaminated (above the maximum level) or is the difference between the sample mean and the maximum level *only* due to random error?

Answer: First, we consider H_0 and H_1 to help in our assessment:

$$H_0: \mu - \mu_0 \quad \text{and} \quad H_1: \mu > \mu_0$$

where μ is the true mean and $\mu_0 = 52$ mg/kg dry weight, the maximum level. A one-tailed test can now be set up. Note that the hypothesis established is one-tailed, since we were looking at the probability that the sample mean was either $>$ or ≤ 52 mg/kg dry weight.

Inserting the values into equation 2.11 reveals that

$$t_{\text{cal}} = \frac{56 - 52}{3.5/\sqrt{7}} = 3.02$$

Referring to Table 2.5, we see that at the 95% confidence level ($p = 0.05$) with $7 - 1 = 6$ DOF, the $t_{\text{table}} = 1.94$. If the $t_{\text{calc}} > t_{\text{table}}$, we reject the null hypothesis. In the case of the investigator, the $t_{\text{cal}} = 3.02 > t_{\text{table}} = 1.94$, thus rejecting H_0 and concluding that the sediment is contaminated above the maximum suggested level.

How can investigators determine whether the difference between the sample mean and population mean is significant? First consider a situation where a large number of results (≥ 30) are obtained, in which case s is considered a good estimate of σ. Here, the z-*test* is an appropriate statistical tool and is calculated as

$$z_{cal} = \frac{\bar{x} - \mu_0}{\sigma/\sqrt{n}} \tag{2.10}$$

The stated null hypothesis in this case would be $H_0: \mu = \mu_0$. H_0 is accepted if the calculated absolute z value is smaller than 1.64 (one-tailed test) or 1.96 (two-tailed test) at the 0.05 level of significance. For a small number of results (≤ 30), we use the t-*test* with $n - 1$ DOF and calculate as

$$t_{cal} = \frac{\bar{x} - \mu_0}{s/\sqrt{n}} \tag{2.11}$$

Consider a situation where investigators are faced with comparing two experimental means (e.g., the difference in the means of two sets of data). Here, for example, one can determine the difference between two sets of replicate measurements or whether two analytical methods give the same values. For the latter, the null hypothesis can be established as $H_0 = \mu_1 = \mu_2$. In most cases, $H_1 = \mu_1 \neq \mu_2$ and the test is a two-tailed version. To test whether $H_0 = \mu_1 = \mu_2$, we can use

$$t_{cal} = \frac{\bar{x}_1 - \bar{x}_2}{s/\sqrt{(1/n_1) + (1/n_2)}} \tag{2.12}$$

where s is the pooled standard deviation and calculated as

$$s^2 = \frac{(n_1 - 1)s_1^2 + (n_2 - 1)s_2^2}{(n_1 + n_2 - 2)} \tag{2.13}$$

For the comparison of two means, t has $n_1 + n_2 - 2$ DOF.

We have just learned how to perform a t-test for the comparison of means, and thus, a way to detect systematic errors that may be present. The question then arises as to how to effectively compare the standard deviations (variances) of two sets of data. In other words, we want a comparison of precision and thus, the random errors in the two datasets. If we assume a normal distribution, an F-test that compares the difference between two sample variances can be used to test $H_0 = \sigma_1^2 = \sigma_2^2$ as

$$F_{cal} = s_1^2/s_2^2 \tag{2.14}$$

with the DOF for the numerator and denominator being $n_1 - 1$ and $n_2 - 1$, respectively. Consider the one-tailed problem presented in Example Problem 2.4.

EXAMPLE PROBLEM 2.4

1. Assume that the nickel concentrations of the sediment samples presented in Example Problem 2.3 were measured by two different methods, each method tested 10 times. The results were as follows:

Method	Mean (mg/kg Dry Weight)	Standard Deviation (mg/kg Dry Weight)
1	57	2.34
2	57	1.21

Determine if s_1^2 (method 1) exceeds s_2^2 (method 2)—a one-tailed characteristic. Our null hypothesis is stated as: H_0: $s_1^2 \le s_2^2$ and H_1: $s_1^2 > s_2^2$. Using Equation 2.14, we have

$$F_{cal} = (2.34)^2/(1.21)^2 = 3.75$$

The table value (Table 2.7) for DOF in each case, and a one-tailed, 95% confidence level is $F_{crit} = 3.179$. Here, $F_{cal} > F_{crit}$, thus we reject H_0, and s_1^2 exceeds s_2^2 significantly.

Suppose now that we wish to consider a two-tailed version of Example Problem 2.4. Here, the null hypothesis is stated as: H_0: $s_1^2 = s_2^2$ and H_1: $s_1^2 \ne s_2^2$. For the F-test, we will now need to refer to Table 2.8, which gives critical values for a two-tailed F at $p = 0.05$. Our F_{cal} remains the same (3.75), but the $F_{crit} = 4.026$. Here, $s_1^2 \ne s_2^2$ and H_0 is rejected.

2.7 OUTLYING RESULTS

Consider the following simulated dataset with six observations generated from a normal population:

$$21.2 \quad 21.4 \quad 21.6 \quad 24.8 \quad 21.1 \quad 21.5$$

Visually, it appears that the value 24.8 is an *outlier*, a result that is outside the range of those typically produced by random errors. To evaluate this value effectively, one needs to incorporate a proper statistical test such as the Dixon's Q-test. This test compares the difference between the suspect value and the nearest value (gap) with the difference between the highest and lowest values (range), producing a ratio calculated from

$$Q_{cal} = \frac{|\text{suspect value} - \text{nearest value}|}{(\text{largest value} - \text{smallest value})} \tag{2.15}$$

Q_{crit} values are found in Table 2.9 for a two-tailed test ($p = 0.05$). Note that if $Q_{cal} > Q_{crit}$, rejection of the suspect value is warranted. From the dataset above, list by increasing value

$$21.1 \quad 21.2 \quad 21.4 \quad 21.5 \quad 21.6 \quad 24.8$$

TABLE 2.7

Critical Values of F for a One-Tailed Test with $p = 0.05$[a]

								DOFs$_1$						
DOFs$_2$	1	2	3	4	5	6	7	8	9	10	12	15	20
1	161.4	199.5	215.7	224.6	230.2	234.0	236.8	238.9	240.5	241.9	243.9	245.9	248.0
2	18.51	19.00	19.16	19.25	19.30	19.33	19.35	19.37	19.38	19.40	19.41	19.43	19.45
3	10.13	9.552	9.277	9.117	9.013	8.941	8.887	8.845	8.812	8.786	8.745	8.703	8.660
4	7.709	6.944	6.591	6.388	6.256	6.163	6.094	6.041	5.999	5.964	5.912	5.858	5.803
5	6.608	5.786	5.409	5.192	5.050	4.950	4.876	4.818	4.772	4.735	4.678	4.619	4.558
6	5.987	5.143	4.757	4.534	4.387	4.284	4.207	4.147	4.099	4.060	4.000	3.938	3.874
7	5.591	4.737	4.347	4.120	3.972	3.866	3.787	3.726	3.677	3.637	3.575	3.511	3.445
8	5.318	4.459	4.066	3.838	3.687	3.581	3.500	3.438	3.388	3.347	3.284	3.218	3.150
9	5.117	4.256	3.863	3.633	3.482	3.374	3.293	3.230	3.179	3.137	3.073	3.006	2.936
10	4.965	4.103	3.708	3.478	3.326	3.217	3.135	3.072	3.020	2.978	2.913	2.845	2.774
11	4.844	3.982	3.587	3.363	3.204	3.095	3.012	2.948	2.896	2.854	2.788	2.719	2.646
12	4.747	3.885	3.490	3.259	3.106	2.996	2.913	2.849	2.796	2.753	2.687	2.617	2.544
13	4.667	3.806	3.411	3.179	3.025	2.915	2.832	2.767	2.714	2.671	2.604	2.533	2.459
14	4.600	3.739	3.344	3.112	2.958	2.848	2.764	2.699	2.646	2.602	2.534	2.463	2.388
15	4.543	3.682	3.287	3.056	2.901	2.790	2.707	2.641	2.588	2.544	2.475	2.403	2.328
16	4.494	3.634	3.239	3.007	2.852	2.741	2.657	2.591	2.544	2.494	2.425	2.352	2.276
17	4.451	3.592	3.197	2.965	2.810	2.699	2.614	2.548	2.494	2.450	2.381	2.308	2.230
18	4.414	3.555	3.160	2.928	2.773	2.661	2.577	2.510	2.456	2.412	2.342	2.269	2.191
19	4.381	3.522	3.127	2.895	2.740	2.628	2.544	2.477	2.423	2.378	2.308	2.234	2.155
20	4.351	3.493	3.098	2.866	2.711	2.599	2.514	2.447	2.393	2.348	2.278	2.203	2.124
30	4.171	3.316	2.922	2.689	2.534	2.421	2.334	2.266	2.211	2.165	2.092	2.015	1.932

[a] DOFs$_1$ = degrees of freedom for numerator; DOFs$_2$ = degrees of freedom for denominator.

TABLE 2.8

Critical Values of F for a Two-Tailed Test with $p = 0.05$[a]

	$DOFs_1$												
$DOFs_2$	1	2	3	4	5	6	7	8	9	10	12	15	20
1	647.8	799.5	864.2	899.6	921.8	937.1	948.2	956.7	963.3	968.6	976.7	984.9	993.1
2	38.51	39.00	39.17	39.25	39.30	39.33	39.36	39.37	39.39	39.40	39.41	39.43	39.45
3	17.44	16.04	15.44	15.10	14.88	14.73	14.62	14.54	14.47	14.42	14.34	14.25	14.17
4	12.22	10.65	9.997	9.605	9.364	9.197	9.074	8.980	8.905	8.844	8.751	8.657	8.560
5	10.01	8.434	7.764	7.388	7.146	6.978	6.853	6.757	6.681	6.619	6.525	6.428	6.329
6	8.813	7.260	6.599	6.227	5.988	5.820	5.695	5.600	5.523	5.461	5.366	5.269	5.168
7	8.073	6.542	5.890	5.523	5.285	5.119	4.995	4.899	4.823	4.761	4.666	4.568	4.467
8	7.571	6.059	5.416	5.053	4.817	4.652	4.529	4.433	4.357	4.295	4.200	4.101	3.999
9	7.209	5.715	5.078	4.718	4.484	4.320	4.197	4.102	4.026	3.964	3.868	3.769	3.667
10	6.937	5.456	4.828	4.468	4.236	4.072	3.950	3.855	3.779	3.717	3.621	3.522	3.419
11	6.724	5.256	4.630	4.275	4.044	3.881	3.759	3.664	3.588	3.526	3.430	3.330	3.226
12	6.554	5.096	4.474	4.121	3.891	3.728	3.607	3.512	3.436	3.374	3.277	3.177	3.073
13	6.414	4.965	4.347	3.996	3.767	3.604	3.483	3.388	3.312	3.250	3.153	3.053	2.948
14	6.298	4.857	4.242	3.892	3.663	3.501	3.380	3.285	3.209	3.147	3.050	2.949	2.844
15	6.200	4.765	4.153	3.804	3.576	3.415	3.293	3.199	3.123	3.060	2.963	2.862	2.756
16	6.115	4.687	4.077	3.729	3.502	3.341	3.219	3.125	3.049	2.986	2.889	2.788	2.681
17	6.042	4.619	4.011	3.665	3.438	3.277	3.156	3.061	2.985	2.922	2.825	2.723	2.616
18	5.978	4.560	3.954	3.608	3.382	3.221	3.100	3.005	2.929	2.866	2.769	2.667	2.559
19	5.922	4.508	3.903	3.559	3.333	3.172	3.051	2.956	2.880	2.817	2.720	2.617	2.509
20	5.871	4.461	3.859	3.515	3.289	3.128	3.007	2.913	2.837	2.774	2.676	2.573	2.464
30	5.567	4.182	3.589	3.245	3.027	2.867	2.746	2.651	2.575	2.511	2.412	2.307	2.195

[a] $DOFs_1$ = degrees of freedom for numerator; $DOFs_2$ = degrees of freedom for denominator.

TABLE 2.9

Critical Values of Q for a Two-Sided Test ($p = 0.05$)

Sample Size	Q_{crit}
3	0.970
4	0.829
5	0.710
6	0.625
7	0.568
8	0.526
9	0.493
10	0.466

Source: From Rorabacker, D. B., *Analytical Chemistry*, 63, 139–146, 1991. With permission from the American Chemical Society.

and calculate Q as

$$Q_{cal} = \frac{|24.8 - 21.6|}{(24.8 - 21.1)} = 0.865$$

Table 2.9 gives a $Q_{crit} = 0.568$ for seven samples. Since $0.865 > 0.568$, the suspect value is rejected. It is in fact an outlier. Remember to report the Q-test and the removal of the suspect data point when preparing the final results for evaluation purposes. Other univariate methods of determining outliers include Grubbs, Walsh, and the range method. Univariate methods (e.g., range method) can also be combined with multivariate principal component analysis (Chapter 6), for example, for enhanced outlier detection.

2.8   ANALYSIS OF VARIANCE

In Section 2.6, we learned how investigators determine whether the difference between the sample mean and population mean is significant. What about determining differences in means among groups or factors for statistical significance? This can be accomplished by performing an analysis of variance (ANOVA). A one-way ANOVA can be used for simple comparisons of treatments or factors. One can also use a two-way ANOVA for combinations of treatments, where two factors (e.g., pH and flow rate), for example, are applied in all possible combinations or levels. These are often termed factorial ANOVA. In Chapter 4, appropriate experimental designs are described and ANOVA models formulated in relation to environmental applications. There are two common applications of ANOVA in environmental analysis presented in this book. The first is for establishing the differences in the results of multi-investigator determinations of an analyte in an environmental sample of interest (see Excel® Spreadsheet Exercise in Section 2.13). The second is for revealing the significance of single-factor effects and interactive effects on an instrumental response when performing experimental design and optimization procedures (see Expanded Research Application in Chapter 4).

TABLE 2.10
ANOVA Summary Table

Source of Variation	Sum of Squares (SS)	DOF	Mean Square (MS)	F
Between Groups (Treatment)	$SS_{treatment} = \sum_{i=1}^{p} \sum_{j=1}^{n_1} (\bar{x}_i - \bar{x})^2$	$p - 1$	$MS_{treatment} = \dfrac{SS_{treatment}}{p-1}$	$F = \dfrac{MS_{treatment}}{MS_{error}}$
Within Groups (Error)	$SS_{error} = \sum_{i=1}^{p} \sum_{j=1}^{n_1} (x_{ij} - \bar{x}_i)^2$	$n - p$	$MS_{error} = \dfrac{SS_{error}}{n-p}$	
Total	$SS_{total} = \sum_{i=1}^{p} \sum_{j=1}^{n_1} (x_{ij} - \bar{x})^2$	$n - 1$		

Essentially, one must consider examining the relationships among variables, or among differences in groups on one or more variables. In ANOVA, the total variance is divided into components—those due to random error (within groups) and those that are due to differences among means (among groups). Subsequently, the latter components are tested for statistical significance using F, and, if significant, H_0 is rejected and we accept the H_1 that the means (population) are in fact different. A generalized ANOVA table with the components required for calculation of significance is presented in Table 2.10. Note that the sum of squares adds up as

$$SS_{treatment} + SS_{error} = SS_{total}$$

Likewise, the DOF also adds up as

$$(p - 1) + (n - p) = n - 1$$

The MS (variances) are found by taking the appropriate SS and dividing it by the DOF The F statistic is calculated by dividing the two MS and then compared to the F_{crit}. As discussed previously, if $F_{cal} > F_{crit}$, the H_0 is rejected.

Suppose that two different investigators measured the boron levels (mg/L) in four samples from a wastewater process. One-way ANOVA can assess whether differences exist between the two investigators' measurements. The mean results for the one-way ANOVA are presented in Table 2.11, with the corresponding means

TABLE 2.11
ANOVA Mean Results for the Investigation of Boron in Wastewater Samples

Level	Number	Mean	Standard Error	Lower 95%	Upper 95%
Investigator 1	4	6.50	0.11319	6.22	6.77
Investigator 2	4	7.07	0.11319	6.79	7.35

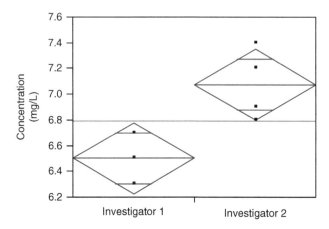

FIGURE 2.6   ANOVA diamond plot with group means and vertical spans representing the 95% confidence intervals.

diamond plot presented in Figure 2.6. The lines across the diamonds represent the group means with the vertical spans representing the 95% confidence intervals. Table 2.12 provides the ANOVA report listing the probability of a difference between the investigators' measurements. Note that the Prob $> F$ (the p-value) is 0.0115, which supports the visual conclusion (Figure 2.6) that there is a significant difference between the investigators.

Related techniques encountered include one-way analysis of covariance (ANCOVA) and one-way multivariate analysis of variance (MANOVA). ANCOVA is similar to ANOVA to the extent that two or more groups are compared on the mean of some dependent variable, and additionally act as controls for a variable (covariate) that may influence the dependent variable. MANOVA is utilized to study two or more related dependent variables simultaneously while controlling for the correlations among the dependent variables. An obvious extension of MANOVA would be the multivariate analysis of covariance (MANCOVA). MANOVA is covered in greater detail in Chapter 6.

TABLE 2.12

ANOVA Report for the Analysis of Boron in Wastewater Samples

Source	DOF	Sum of Squares	Mean Square	F-Ratio	Prob > F
Investigators	1	0.66125	0.66125	12.9024	0.0115
Error	6	0.30750	0.05215	—	—
C. Total	7	0.96875	—	—	—

2.9 REGRESSION AND CALIBRATION METHODS

One of the most frequently used statistical methods utilized in *calibration* is *linear regression*. In its simplest terms, regression describes a group of methods that summarize the degree of association between one variable (or set of variables) and another variable (or set of variables). The most common method used in this process is least squares linear regression, which finds the best curve through the data that minimizes the sums of squares of the residuals. Residuals are calculated by taking the difference between the predicted value and the actual value. Other least squares regression methods are the logarithmic, exponential, and power models. Various alternative regression methods are available for use in investigations and include multiple linear regression, ranked regression, principal component regression, non-linear regression, and partial least squares regression. The majority of these are covered in detail in Chapter 6.

Consider a calibration process of measuring the instrument response (y) of an analytical method to known concentrations of analyte (x) using model building and validation procedures. These measurements, along with the predetermined analyte concentrations, encompass a calibration set. This set is then used to develop a mathematical model that relates the amount of the sample to the instrumental measurements. Here we develop what is commonly called a *calibration curve*, using the method of least squares regression to find the best straight line through the calibration set. The equation of the straight line is

$$y = mx + b \tag{2.16}$$

where m is the slope of the line and b is the y-intercept (point at which the line crosses the y-axis). Upon the determination of the slope and intercept, the concentration (x-value) corresponding to the relevant instrument measurement signals (y-value) can be calculated by a rearrangement of Equation 2.16.

Which way to go?

A one-way ANOVA is used when a single factor is involved with several levels and multiple observations at each level.

A two-way ANOVA is used to study the effects of two factors separately (main effects) and together (interactive effects).

The method of least squares finds the best-fit line by adjusting the line to minimize the vertical deviations between the points on it (Harris 2005). Note that uncertainties arise in both m and b, and are related to the uncertainty in measuring each value of y. Additionally, instrument signals derived from any test material are subject to random errors. Typically, goodness-of-fit measures for calibrations involve determining the *correlation coefficient* (r) or the *coefficient of determination* (r^2)— the fraction of the variance in one variable that is explained by the linear model in the other. An assumption of the latter is that the variables are normally distributed and that the r^2 value is sensitive to outliers (see following discussion).

A *blank* is an analytical control consisting of all reagents, internal standards, and surrogate standards, which are carried through the entire analytical procedure. There are a variety of blanks (e.g., method, field, reagent) utilized in environmental analysis.

There are those who believe that the correlation coefficient is potentially misleading and should be avoided when describing goodness-of-fit in a linear model since the values of x are not random quantities in a calibration experiment (Analytical Methods Committee 1994). In addition, one cannot reasonably assume that a dataset with $r = 0.99$ (where $r = 1$ is perfect positive correlation) has greater linearity than a set with $r = 0.95$. Other statistical tests (e.g., lack-of-fit, Mandel's fitting test) are argued to be more suitable for the validation of the linear model (Van Loco et al. 2002). Relative standard deviation has prevalent acceptance as a measure of the error for an average calibration with the relative standard error (RSE) providing an indicator of error to any form of a weighted regression function. Weighted regression reflects the behavior of the random errors and can be used with functions that are both linear and nonlinear in nature. This process incorporates the nonnegative constants (weights) associated with each data point into the curve-fitting criterion. Ultimately, the size of the weight is indicative of the precision of the information contained in the associated observation. It is important to remember, though, that weighted regression functions are also sensitive to the presence of outliers.

Uncertainty can be quantified by displaying confidence limits around the best-fit line. With multiple replicates of each standard, error bars (e.g., $\pm 3\,\sigma$) can be added to the mean values plotted. Consider the example of a calibration graph in Figure 2.7a, which displays the regression equation, the coefficient of determination, fit mean, linear regression fit, and 95% confidence limits. Figure 2.7b displays the residuals plot.

Consider an environmental application study by Barco et al. (2006), who performed a series of three calibrations for phosphite P(+III) determination in ultrapure water using an automated flow injection technique. As shown in Table 2.13, reproducibility for replicate injections of P(+III) standards (0–25 µM) was typically <3.9% RSD ($n = 3$). The pooled data also showed good linear correlation ($r^2 = 0.99$). An example of a calibration curve is shown in Figure 2.8.

In this section, we have considered traditional univariate calibration techniques. In an ideal chemical measurement using high-precision instrumentation, an investigator may obtain selective measurements linearly related to analyte concentration. However, univariate techniques are very sensitive to the presence of outlier points in the data used to fit a particular model under normal experimental conditions. Often, only one or two outliers can seriously skew the results of a least squares analysis.

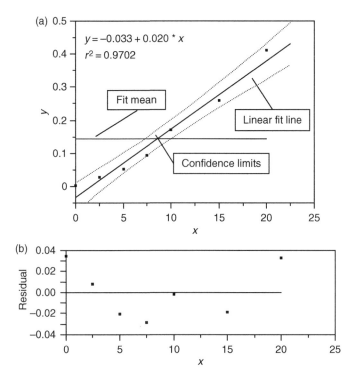

FIGURE 2.7 (a) Calibration graph displaying the regression equation, the coefficient of determination, fit mean, linear regression fit, and 95% confidence limits; and (b) the residual plot.

The problems of *selectivity* and interferences (chemical and physical) also limit the effectiveness of univariate calibration methods and cause some degree of nonlinearity. In addition, such calibration techniques are not appropriate for the multitude of data collected through the sensitive, high-throughput instrumentation currently used in environmental analyses. These datasets often contain large amounts of information, but in order to extract this information fully and interpret it correctly, methods incorporating a multivariate approach are needed. Multivariate calibration and analysis techniques are discussed in Chapter 6.

2.10 SENSITIVITY AND LIMIT OF DETECTION

Novice investigators often confuse the terms *sensitivity* and *limit of detection* (LOD) of a technique. Sensitivity (calibration sensitivity) is defined as the change in the response signal to changes in analyte concentration. In other words, sensitivity is the slope of the calibration curve represented by

$$m = \frac{\Delta y}{\Delta x} \tag{2.17}$$

TABLE 2.13

Calibration Data for the Determination of Phosphite in Ultrapure Water Samples

[P(III)] (μM)	Calibration 1		Calibration 2		Calibration 3		Pooled Data	
	Mean Absorbance	RSD (%) ($n=3$)	Mean Absorbance	RSD (%) ($n=3$)	Mean Absorbance	RSD (%) ($n=3$)	Mean Absorbance	RSD (%) ($n=3$)
0	0.0004	0.8	0.0005	1.4	0.0005	1.2	0.0005	12.3
5	0.0110	5.9	0.0125	3.9	0.0155	3.4	0.0131	7.2
10	0.0182	1.3	0.0181	0.9	0.0179	6.2	0.0181	0.8
15	0.0295	2.2	0.0302	1.8	0.0305	1.8	0.0301	1.7
25	0.0532	3.4	0.0529	3.1	0.0532	2.2	0.0530	0.4
r^2	0.9935		0.9914		0.9924		0.9921	
Slope	0.0021	0.0021	0.0021	0.0021	0.0021	0.0021	0.0021	0.0021
Intercept (absorbance)	−0.0006	−0.0006	0.0002	0.0002	−0.0006	−0.0006	0.0002	0.0002

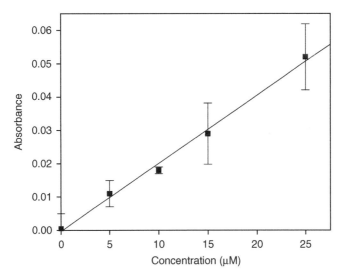

FIGURE 2.8 Representative phosphite calibration curve. Error bars = 3σ. (From Barco et al., *Talanta*, 69, 1292–1299, 2006. With permission from Elsevier.)

The LOD for an analytical method is the lowest concentration of an analyte that gives an instrument signal distinguishable (within a certain level of confidence) from the blank signal. If we consider methods that employ a calibration curve, the LOD can be determined as

$$\text{LOD} = \frac{ks_b}{m} \tag{2.18}$$

where k is the numerical factor based on the desired confidence level, s_b the standard deviation of the blank measurements, and m is the calibration sensitivity. Typically, $k = 3$, which corresponds to a confidence level of 98.3%, is chosen (Skoog et al. 2004). As discussed by various authorities (IUPAC 1997, Miller and Miller 2005), there is an increasing trend to define the LOD as

$$\text{LOD} = y_b + 3s_b \tag{2.19}$$

where y_b is the mean of the blank signals. Note that in both cases, the LOD must be based on multiple measurements of a *blank*—a hypothetical sample containing zero variable concentration (IUPAC 1997).

One will encounter various terms and phrases that describe the methods to derive the detection limits. The method detection limit (MDL), for example, is the lowest concentration of an analyte that a method can detect reliably and is statistically different from a blank carried through the complete method (excluding extraction and sample pretreatment). The MDL is based on replicate measurements with a specific confidence level (typically 95% or 99%, Section 2.5). Another term that is used is the instrument detection limit (IDL), which is the concentration equivalent to the smallest signal (arising from the analyte of interest) that can be distinguished from background noise by a particular instrument.

2.11 BAYESIAN STATISTICS CONSIDERED

We have just learned the classical (frequentist) viewpoint of statistical theory—one in which probabilities are only associated with data taken from repeated observations. These methods, although commonly used and highly praised in environmental analysis, can produce uncertainty, especially with regard to risk assessment where the inputs in an exposure model are not always precisely known (Bates et al. 2003). While the major emphasis of this book is on the analysis and interpretation of environmental data, mention must also be made of the developing trend to incorporate Bayesian statistics to make inferences from classical models while accounting for uncertainty in the inputs to models.

In contrast to the classical approach, Bayesian statistical inference requires the assignment of prior probabilities; those based on existing information (e.g., previously collected data, expert judgment), to experimental outcomes. Regardless of the sample size, these results can then be used to compute posterior probabilities of original hypotheses based on the data available. Whether used separately or in conjunction with classical methods, Bayesian methods provide environmental scientists with a formal set of tools for analyzing complex sets of data. Distinct advantages of such techniques include (Uusitalo 2007):

1. suitability for use when missing data is evident;
2. allowing a combination of data with domain knowledge;
3. facilitating the study of casual relationships among variables;
4. providing a method for avoiding overfitting of data;
5. showing good prediction accuracy regardless of sample size;
6. possessing the capability to be combined with decision analytic tools for aiding management-related activities.

The use of Bayesian modeling techniques in environmental analysis is becoming increasingly evident as highlighted by a variety of current research applications (Bates et al. 2003, Qian et al. 2003, Khali et al. 2005, Englehart and Swartout 2006, Cressie et al. 2007). A closer examination of Qian et al. (2003) reveals how Bayesian hierarchical methods can be used for the detection of environmental thresholds to examine changes of selected populations, communities, or ecosystem attributes along a gradient of environmental conditions. Most traditional statistical techniques are neither suitable for estimating such thresholds, nor adequate for estimating uncertainty in their predictions (Qian et al. 2003).

In the Qian study, a nonparametric method was compared with a Bayesian model based on the change in the response variable distribution parameters. Both methods were tested using macroinvertebrate composition data from a mesocosm experiment conducted in the Florida Everglades where phosphorus is the *limiting nutrient*.

A *limiting nutrient* is a chemical necessary for plant growth, but often only available in a concentration insufficient to support sustained growth. If a limiting nutrient is introduced to a system, biological growth will be promoted until limitation by the lack of that or another limiting factor occurs. Typically, nitrogen and phosphorus are limiting nutrients in marine and freshwater systems, respectively.

The investigators used the percent of phosphorus-tolerant species and a dissimilarity index as the response variable to produce a defined total phosphorus concentration threshold. The CDF plot (Figure 2.9) was then used to determine the probability of exceeding the threshold level. The investigators do note, however, two minor limitations of the Bayesian method. First, it requires specific information on the distribution of the response variable. Second, computation with the Bayesian method can be more intensive in nature than with the nonparametric method.

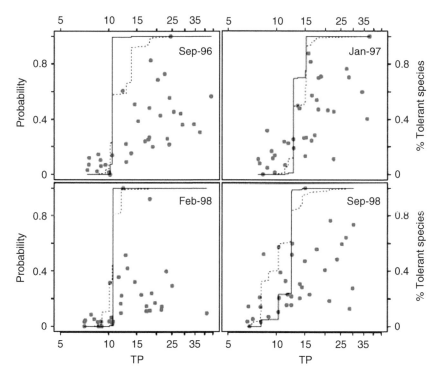

FIGURE 2.9 Cumulative distributions estimated for percent phosphorus tolerant species using the nonparametric (dashed line) and the Bayesian (solid line) methods. Risk of exceeding the TP threshold is presented. (From Qian, S. S., King, R. S., and Richardson, C. J., *Ecological Modelling*, 166, 87–97, 2003. With permission from Elsevier.)

2.12 EXPANDED RESEARCH APPLICATION I—STATISTICAL MERITS OF CALCULATING TRANSFER COEFFICIENTS AMONG ENVIRONMENTAL MEDIA

As discussed earlier, statistical methods generally assume that a normal distribution applies to given datasets. Does the same apply to the complex sets of data obtained from environmental analyses? To answer this question, consider a paper by Juan et al. (2002) who explored the statistical merits of datasets from environmental pollution studies, with particular reference to transfer coefficients (TCs) and input–output balance values of PCBs in animals and humans. TCs are essentially ratios describing the movement of chemicals from one environmental or biological compartment to another. Consider the study described here.

2.12.1 STATEMENT OF PROBLEM

It is highly desirable to base TCs on measurements from a large population of samples or individuals. However, concentrations in environmental media have been shown to have a log-normal distribution; that is, the arithmetic means and standard deviations of measurements of X and Y incorrectly describe the distribution of the data used to calculate TCs (Sielken 1987, Juan et al. 2002). This can ultimately lead to biased TC estimates and invalid confidence intervals. Table 2.14 lists the TC estimators and formulae considered in this study.

2.12.2 RESEARCH OBJECTIVES

To investigate possible statistical methods of calculating TCs and properly assessing bias-associated errors using each method considering both normal and log-normal distributions.

TABLE 2.14
TC Estimators and Formulas

TC Estimator	Description	Formula Used in Estimation
TC_{RM}	Ratio of means	$\hat{b}_{RM} = \dfrac{\bar{y}}{\bar{x}}$
TC_{LS}	Least squares	$\hat{b}_{LS} = \dfrac{\sum x_i y_i}{\sum x_i^2}$
TC_{MR}	Mean of ratios	$\hat{b}_{MR} = \dfrac{1}{n} \sum \dfrac{y_i}{x_i}$
TC_{GMR}	Geometric mean regression	$\hat{b}_{GMR} = \left(\dfrac{\sum y_i^2}{\sum x_i^2} \right)^{1/2}$
TC_{RGM}	Ratio of geometric means	$\hat{b}_{RGM} = \exp(\bar{y} - \bar{x})$
TC_{CRGM}	Corrected ratio of geometric means	$\hat{b}_{CRGM} = \hat{b}_{RGM} \exp\left(-\dfrac{1}{2} S_D^2/n\right)$

Source: Modified from Juan, C.-Y., Green, M., and Thomas, G. O., *Chemosphere*, 46, 1091–1097, 2002. With permission from Elsevier.

2.12.3 EXPERIMENTAL METHODS AND CALCULATIONS

In a normal distribution, calculation of TCs generally involves on the formula (Juan et al. 2002)

$$TC = \frac{Y}{X} \tag{2.20}$$

where X and Y are chemical concentration measurements in two different media. To correctly account for errors before the calculation of TCs, the investigators must first estimate the standard deviations as (Juan et al. 2002)

$$\left(\frac{\sigma_{TC}}{\mu_{TC}}\right)^2 = \left(\frac{\sigma_Y}{\mu_Y}\right)^2 + \left(\frac{\sigma_X}{\mu_X}\right)^2 - 2r_{YX}\left(\frac{\sigma_Y}{\mu_Y}\frac{\sigma_X}{\sigma_Y}\right) \tag{2.21}$$

where r_{YX} is the correlation between Y and X. The investigators considered producing a TC for each sampling occasion and calculated the arithmetic mean of all derived TCs (TC_{MR}). Another approach considered was to combine the least-squares estimates (TC_{LS}) of the slopes derived from the regression of Y on X, and that of X on Y—an estimate of TC_{GMR}.

For a log-normal distribution, calculations of TCs take on the formula (Juan et al. 2002)

$$\log TC = (\log Y) - (\log X) \tag{2.22}$$

To produce the TC_{RGM}, the mean of the log TC values are taken and then transformed back to produce an average TC as the ratio of the geometric means of measured Y and X. Transformation of log TC to the pure TC results in a log-normal distribution (and thus, biased in nature). To correct this, the investigators subtracted the bias from TC_{RGM} to give the corrected ratio of geometric means (TC_{CRGM}). Here, differences in log Y and log X are given by

$$D = \log Y - \log X \tag{2.23}$$

with the sample variances of these differences, S_D^2, computed. They then estimated the accuracy of the estimated TCs using

$$\{\exp(S_D^2/n) - 1\}TC_{CRGM}^2 \tag{2.24}$$

2.12.4 RESULTS AND INTERPRETATION

In Section 2.3, we learned that as sample size increases, the sampling distribution of sample means resembles a normal distribution. If we consider TC_{RM} and calculate the standard error approximated using Equation 2.21, there is an assumption that both the numerator and denominator demonstrate a log-normal distribution. The investigators stated, however, that as sample size increases, $\bar{y}$ and $\bar{x}$ will have a distribution leaning

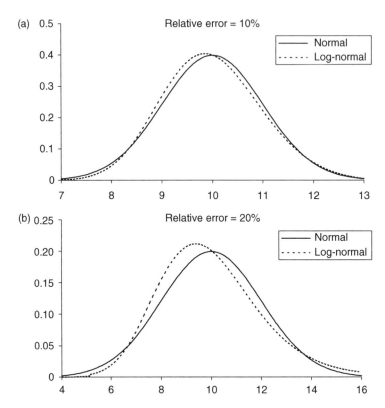

FIGURE 2.10   Plot considering both normal and log-normal distributions with (a) relative error = 10%; and (b) relative error = 20% in the study of the statistical calculation of TCs (From Juan, C.-Y., Green, M., and Thomas, G. O., *Chemosphere*, 46, 1091–1097, 2002. With permission from Elsevier.)

toward normal. To test the influence of the size of relative errors, a plot was constructed taking into consideration both normal and log-normal distributions (Figure 2.10). The x-axis includes nominal values. The y-axis values are $f(x)$, produced using a probability function equation, to represent the distributions with the parameters specified. These values are also nominal, but the proportional differences seen between the y-axis in Figures 2.10a and 2.10b relate to actual proportional differences in the probability distribution heights.

To better understand Figure 2.10, and to examine the accuracy of this approximation obtained using Equation 2.21, a simulation study based on assuming $\bar{y}$ and $\bar{x}$ having a normal distribution was initiated. Table 2.15 shows the results for various values of $\bar{y}$ and $\bar{x}$ (variance calculated using Equation 2.20 with $r_{yx} = 1$) compared to the variance calculated using Equation 2.23. As shown, approximation with Equation 2.21 is accurate when the relative errors are both <10%, although reportedly sensitive to the size of the relative error in $\bar{x}$. When approaching 20%, the variance approximation becomes less accurate.

TABLE 2.15

The Accuracy of Variance Approximation Using Equations 2.21 and 2.24

Relative Error in $\bar{y}$ (%)	Relative Error in $\bar{x}$ (%)	Bias in Approximation (%)
10	10	-1.6
20	20	-8.0
10	20	-15.8

Source: Modified from Juan, C.-Y., Green, M., and Thomas, G. O., *Chemosphere*, 46, 1091–1097, 2002. With permission from Elsevier.

Next, simulation studies using each estimator were performed on sample sizes of $n = 10$ and $n = 40$ with full results presented in Table 2.16. As can be seen, TC_{CRGM} has the least amount of bias, particularly when $n = 40$. The standard error for TC_{CRGM} is also the smallest, indicating greater efficiency in estimation. All the other estimators have a particular amount of bias, especially TC_{LS} (-9%) and TC_{MR} (17%). Is it appropriate to assume the sample size reduced the amount of bias? According to their results, yes, but with only a slight decrease in bias (TC_{RM}, TCG_{MR}, TC_{RGM}) when the sample size is increased to $n = 40$.

Finally, validation of the methods used in the study with real PCB samples was performed. Here, PCB input–output balances in humans measured from daily food intake and daily output (analysis of faeces) were calculated as

$$\text{Input--output balance} = \frac{\text{daily output flux}}{\text{daily input flux}} \tag{2.25}$$

where the flux was measured in ng/day. As shown in Table 2.17, TC_{RM} and TC_{CRGM} give similar values for the input–output balances for two separate PCB samples. The

TABLE 2.16

Bias and Standard Error for Each TC Estimator on a Simulated Set of Data with Moderate Measurement Error (S.D. = 0.4 × mean) in Both X and Y

Estimator	Bias (%) $n = 10$	Standard Error	Bias (%) $n = 40$	Standard Error
TC_{RM}	2	0.162	0.5	0.080
TC_{LS}	-9	0.180	-13	0.097
TC_{MR}	17	0.182	17	0.091
TC_{GMR}	3	0.200	0.9	0.108
TC_{RGM}	1.6	0.146	0.4	0.072
TC_{CRGM}	-0.05	0.144	-0.02	0.071

Source: Modified from Juan, C.-Y., Green, M., and Thomas, G. O., *Chemosphere*, 46, 1091–1097, 2002. With permission from Elsevier.

TABLE 2.17

TC (Input–Output Balance) Estimates for Samples PCB153 and PCB138

Estimator	PCB153	PCB138
TC_{RM}	0.765	0.932
TC_{LS}	0.655	0.754
TC_{MR}	0.907	1.008
TC_{GMR}	0.922	1.238
TC_{RGM}	0.808	1.023
TC_{CRGM}	0.791	0.992

Source: Modified from Juan, C.-Y., Green, M., and Thomas, G. O., *Chemosphere*, 46, 1091–1097, 2002. With permission from Elsevier.

TABLE 2.18

TC_{RM} and TC_{CRGM} Standard Error Calculations

Parameter	PCB153	PCB138
Relative error $\bar{y}$	10%	10%
Relative error $\bar{x}$	16%	18%
Standard error TC_{RM}	0.0219	0.0422
Standard error TC_{CRGM}	0.0261	0.0623

Source: Modified from Juan, C.-Y., Green, M., and Thomas, G. O., *Chemosphere*, 46, 1091–1097, 2002. With permission from Elsevier.

other estimators vary in greater detail. Standard errors for both TC_{RM} and TC_{CRGM} are shown in Table 2.18. Here, the investigators concluded that the level of relative error in $\bar{x}$ suggests that the estimated errors for TC_{RM} are underestimations in both cases.

2.12.5 SUMMARY AND SIGNIFICANCE

In summary, the investigators concluded that the ideal method for estimating TCs for PCB input–output studies was the incorporation of TC_{CRGM}. However, they advise caution when using TC_{RM}, although it has been shown to be useful, because of the underestimation of standard error when the measurement error on X is larger than that on Y. The study proved the power of using combined factor estimators for the treatment and evaluation of complex environmental data with a log-normal distribution.

2.13 INTRODUCTION TO EXCEL

Excel® is one of today's most widely used spreadsheet tools and a valuable resource in introductory-level statistics. Its basic nature and ease of application make it popular with beginners, although higher-level computing tools such as JMP®, MiniTab®, and MATLAB® are favored by more advanced users. The intent of this section is to introduce the basic concepts of Excel® and to show the related applications. More extensive explanations of Excel® can be found in dedicated sources (Frye 2003, Walkenback 2007).

The beauty of Excel® is that everything is built in a spreadsheet program with rows and columns, formatting controls, and mathematical functions ready to use. (See Appendix I for common shortcut and combination keys.) A very attractive addition to Excel® is the Analysis ToolPak add-in that allows custom commands and procedures that are useful in performing statistical analyses. Consider a few example applications given here showing the spreadsheet and related functions. Detailed steps and visual manipulations are provided to assist learning.

Excel® Exercise 2.1. Perform a spreadsheet analysis of Example Problem 2.1 in Section 2.3. Use the spreadsheet functions to see how results compare with manual calculations.

> **Step 1:** Enter the measurements in a column next to the random sampling stations, click on the **Function** tab, and move curser to **Average** (see Figure 2.11).
>
> **Step 2:** Next, highlight the cells of interest and click **OK**. Alternatively, you can enter the average formula = **AVERAGE(C9:C13)** manually and click **Enter** (see Figure 2.12).
>
> **Step 3:** The mean value is entered and we can now proceed to calculate standard deviation. Click again on the **Function** tab, then on **STDEV** and highlight the values of interest (see Figure 2.13).
>
> **Step 4:** Click **OK** and the desired information will be listed (see Figure 2.14).

Excel® Exercise 2.2. Four analysts performed five measurements each of the lead concentration (mg/L) of a freshly collected wastewater sample (Table 2.19). Perform a one-way ANOVA to determine if there is a significant difference among the analysts' measurements.

> **Step 1:** Select the **Data Analysis** tab under **Tools**, then select **ANOVA: Single Factor** and click **OK**. Next, choose the Input Range in the window that appears and highlight cells C7 through F11. Note that an α value of 0.05 was chosen in the options window (see Figure 2.15).
>
> **Step 2:** Click **OK** and the ANOVA results will appear in a separate worksheet (see Figure 2.16).

Here, Excel® automatically displays F_{crit} corresponding to the α level chosen earlier. Since F is greater than F_{crit}, we can safely conclude that the measurements differ significantly among analysts.

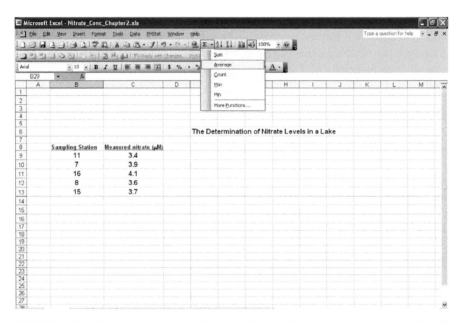

FIGURE 2.11 Entering random environmental sampling station data into an Excel® spreadsheet.

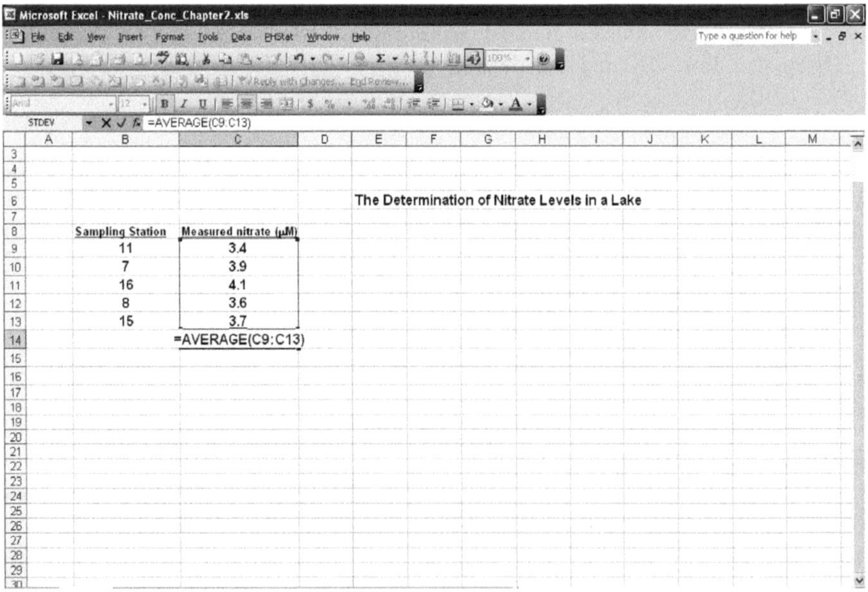

FIGURE 2.12 Performing the Excel® AVERAGE function on the sampling station data.

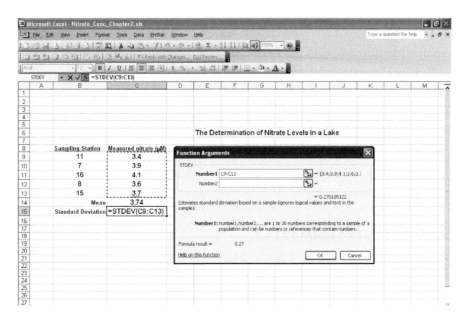

FIGURE 2.13   Selection of values for subsequent Excel® STDEV determination.

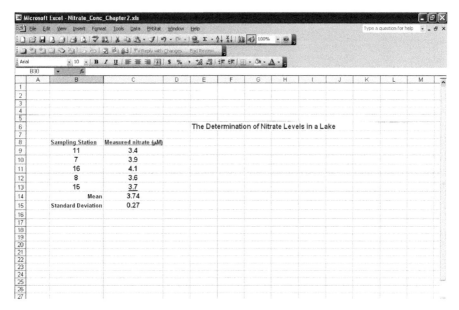

FIGURE 2.14   Desired average and standard deviation values listed on the final Excel® spreadsheet.

TABLE 2.19

Analyst Measurements of the Lead Concentration (mg/L) of a Freshly Collected Wastewater Sample

Measurements	Analyst 1	Analyst 2	Analyst 3	Analyst 4
1	12.2	11.6	12.9	11.4
2	11.8	10.9	13.1	10.9
3	12.6	11.8	12.0	11.2
4	12.4	11.5	11.9	10.9
5	11.9	12.0	12.1	10.9

2.14   CHAPTER SUMMARY

Proper sampling and statistical data analysis methods are vital in any environmental monitoring campaign, especially with regard to modern efforts where sensitive, high-throughput instrumentation is used to generate large amounts of data for regulatory and compliance purposes. This chapter covered important statistical considerations and tests to evaluate effectively both the qualitative and quantitative components of the environmental analysis process including sampling design, analysis, data interpretation and reporting. A multitude of statistical analyses are available for the wide range of concerns in each environmental matrix. Although no single statistical test is

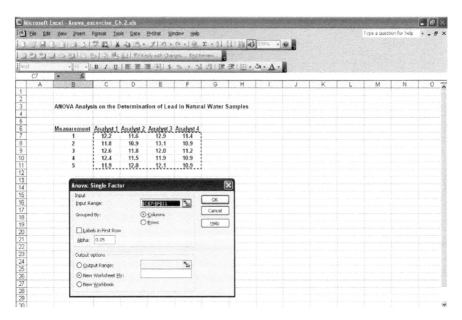

FIGURE 2.15   Data from the four analysts entered as input range values for subsequent Excel® ANOVA determination.

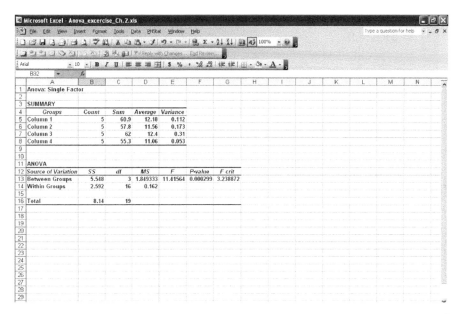

FIGURE 2.16 Full ANOVA results from the four analysts displayed on the final Excel®
spreadsheet.

appropriate for all the various issues, students and investigators can still utilize the
information provided in this chapter in order to obtain reliable and effective interpretation of environmental data.

2.15   END OF CHAPTER PROBLEMS

1. Describe the differences between a population and a sample.
2. What are the differences between accuracy and precision? How do we test
   for each?
3. What are the characteristics of a normal distribution?
4. Describe the differences between parametric and nonparametric statistical
   methods. How do investigators know when to use nonparametric methods?
5. Measurements of the iron concentration in 50 water samples taken from a
   lake showed a mean of 0.734 mg/L and a standard deviation of 0.038 mg/L.
   Determine the 95% confidence limits for the mean concentration.
6. The results from monitoring benzene concentrations (mg/L) in two groundwater wells are as follows

Measurement	Well 1	Well 2
1	3.35	4.99
2	2.25	3.08
3	2.09	2.99
4	2.01	3.16
5	2.15	3.15

Calculate the mean and standard deviation from the replicate datasets given for each well.

7. Consider the data obtained from a multidetermination experiment of total copper (μM) in river water:

$$70.10, 69.62, 69.70, 69.65$$

a. Determine whether any of the data should be rejected at 90% confidence.

8. You have just performed a phosphate PO_4^{3-} spectrophotometric calibration study to help determine the unknown PO_4^{3-} concentration in a natural water sample. The relevant calibration data is given in tabular form as

PO_4^{3-} Standard (μM)	Mean Absorbance ($n = 3$)
0	0.0001
2.5	0.0137
5	0.0278
10	0.0500
15	0.0780

a. Use the Excel® Analysis ToolPak Regression Tool to perform a linear least-squares analysis of your data. Be sure to include the table highlighting the Summary Output.

b. Once plotted, use Excel® functions to find the unknown PO_4^{3-} concentration of a sample giving an absorbance of 0.0194.

9. Two investigators analyze the same wastewater sample for the dissolved O_2 (DO) level. Investigator A makes six determinations and obtains $s_A = 0.14$, whereas investigator B makes five determinations and obtains $s_B = 0.05$. Use the appropriate statistical test and table to determine if there are significant differences in the results at the 95% confidence level.

10. Four investigators performed three determinations of nitrate in water using the same procedure. The results (in μM) were

Investigator	1	2	3	4
	6.7	6.3	6.9	7.0
	6.9	6.2	7.0	7.2
	6.8	6.1	7.1	7.1

a. Determine whether there is a significant difference ($p = 0.05$) in the mean nitrate value measured by each investigator.

b. Use Excel® to calculate and present results in a proper ANOVA table format.

REFERENCES

Analytical Methods Committee. 1994. Is my calibration linear? *Analyst* 119: 2363–2365.

Barco, R. A., Patil, D. G., Xu, W., Ke, L., Khachikian, C. S., Hanrahan, G., and Salmassi, T. M. 2006. The development of iodide-based methods for batch and on-line determinations of phosphite in aqueous samples. *Talanta* 69: 1292–1299.

Bates, S. C., Cullen, A., and Raftery, A. E. 2003. Bayesian uncertainty assessment in multi-component deterministic simulation models for environmental risk assessment. *Environmetrics* 14: 355–371.

Burns, D. T., Danzer, K., and Townshend, A. 2002. Use of terms "recovery" and "apparent recovery" in analytical procedures. *Pure and Applied Chemistry* 74: 2201–2205.

Cressie, N., Buxton, B. E., Calder, C.A., Graigmile, P. F., Dong, C., McMillian, N. J., Morara, M., Santner, T. J., Wang, K., Young, G., and Zhu, J. 2007. From sources to biomarkers: a Bayesian approach to human exposure modeling. *Journal of Statistical Planning and Inference* 137: 3361–3379.

Englehart, J. D. and Swartout, J. 2006. Predictive Bayesian microbial dose-response assessment based on suggested self-organization in primary illness response: *Cryptosporidium parvum*. *Risk Analysis* 26: 543–554.

Frye, C. 2003. *Microsoft Excel 2003 Step by Step*. New York: Microsoft Press.

Gy, P. 2004a. Sampling of discrete materials—a new introduction to the theory of sampling. I. Qualitative approach. *Chemometrics and Intelligent Laboratory Systems* 74: 7–24.

Gy, P. 2004b. Sampling of discrete materials. II. Quantitative approach—sampling of zero-dimensional objects. *Chemometrics and Intelligent Laboratory Systems* 74: 25–38.

Harris, D. C. 2005. *Exploring Chemical Analysis*, 3rd Edition. New York: W.H. Freeman and Company.

Haswell, S. J. 1992. *Practical Guide to Chemometrics*. New York: Marcel Dekker.

Juan, C.-Y., Green, M., and Thomas, G. O. 2002. The statistical merits of various methods of calculating transfer coefficients between environmental data—development of the ideal formula for data-sets with a log-normal distribution. *Chemosphere* 46: 1091–1097.

Khalil, A., McKee, M., Kemblowski, M., and Asefa, T. 2005. Sparse Bayesian learning machine for real-time management of reservoir releases. *Water Resources Research* 41: 1–15.

Ingamells, C. O. and Switzer, P. 1973. A proposed sampling constant for use in geochemical analysis. *Talanta* 20: 547–568.

IUPAC. 1997. *Compendium of Analytical Nomenclature*, 3rd Edition, Research Triangle Park: International Union of Pure and Applied Chemistry (IUPAC).

Miller, J. N. and Miller, J. C. 2005. *Statistics and Chemometrics for Analytical Chemistry*. Essex: Pearson Education Limited.

Qian, S. S., King, R. S., and Richardson, C. J. 2003. Two statistical methods for the detection of environmental thresholds. *Ecological Modelling* 166: 87–97.

Ramsey, M. H. 2002. Appropriate rather than representative sampling, based on acceptable levels of uncertainty. *Accreditation and Quality Assurance* 7: 274–280.

Rorabacker, D. B. 1991. Statistical treatment for rejection of deviant values: critical values of Dixon's "Q" parameter and related subrange ratios at the 95% confidence level. *Analytical Chemistry* 63: 139–146.

Sielken, R. L. 1987. Statistical evaluations reflecting skewness in the distribution of TCDD levels in human adipose tissue. *Chemosphere* 16: 2135–2140.

Skoog, D. A., West, D. M., Holler, F. J., and Crouch, S. R. 2004. *Fundamentals of Analytical Chemistry*. Belmont, CA: Brooks/Cole.

Thompson, M. 1998. Uncertainty of sampling in chemical analysis. *Accreditation and Quality Assurance* 3: 117–121.

U.S. EPA. 1997. *Guidance for Data Quality Assessment Practical Method for Data Analysis.* EPA/QA/G9. Washington D.C.: US Environmental Protection Agency.

Uusitalo, L. 2007. Advantages and challenges of Bayesian networks in environmental modelling. *Ecological Modelling* 203: 312–318.

Van Loco, J., Elskens, M., Croux, C., and Beemaert, H. 2002. Linearity of calibration curves: use and misuse of the correlation coefficient. *Accreditation and Quality Assurance* 7: 281–285.

Walkenback, J. 2007. *Excel 2007 Bible*. New York: Wiley.

3 Quality Assurance in Environmental Analysis

3.1 THE ROLE OF CHEMOMETRICS IN QUALITY ASSURANCE

Environmental monitoring programs, especially those involving complex matrices, speciation requirements, and determination at ultratrace concentrations, are vulnerable to errors and discrepancies in the sampling, storage, and analysis stages. *Quality assurance* (QA) for such investigations, including sample collection, preparation, storage, and analysis, is crucial for providing scientifically reliable and legally defensible data from which legislative decisions can be effectively generated and implemented. QA ensures that *quality control* (QC) takes a twofold approach in that it requires both monitoring of the QA process being studied and elimination of the causes of substandard performance.

Chemometric techniques can be utilized in the analysis of analytical instrument calibration data, evaluation of measurement uncertainty, prediction, validation, classification, and, as we learn in Chapter 4, in the proper design of experiments. A study by Ortiz-Estarelles et al. (2001), for example, examined the use of multivariate statistical process control (MSPC) tools for water quality estimation. Such tools are based on principal component analysis (PCA) and partial least-squares (PLS) models adapted for the detection of errors in methods employed for routine analysis. More detailed discussions on these, and other multivariate analysis techniques, are presented in Chapter 6. Also note that there is an extended research application in Section 3.13.

3.2 QUALITY ASSURANCE CONSIDERATIONS

A variety of international committees and regulatory bodies are actively working on developing, maintaining, and enforcing standardization of laboratory methods and accreditations for a myriad of sectors (e.g., analytical testing, environmental, food, and pharmaceutical industries). Presented here are a few items to consider in the initial stages of the development of any QA management plan.

3.2.1 PROJECT PLANNING AND PREPARATION

Like any management plan, environmental monitoring and *data collection* projects require a systematic planning approach to guide the proposed project effectively and to help ensure quality data. Let us now consider the United States Environmental Protection Agency (U.S. EPA)-developed data quality objectives (DQO) process, which we touched upon in Chapter 2, in much greater detail. Outlined here are the seven generalized steps of the DQO process, each providing valuable output information for subsequent steps:

1. *Statement of the problem*—definition of the problem, identifying planning team and strategies, and formulating a defined budget and schedule;
2. *Identifying project decisions*—identifying study questions and defining alternative actions;
3. *Identifying key decision inputs*—identifying types of information required for the decision process; identifying action levels and vital sampling, storage, and analysis methods;
4. *Defining study boundaries*—defining sample characteristics, spatial and temporal limits, variability, and units for proper decision-making;
5. *Developing a decision rule*—defining statistical parameters, action levels, and the development of a decision rule that links parameters and action levels to aid decision-makers;
6. *Specifying tolerable limits on decision errors*—the establishment of acceptable limits for decision errors in relation to health effects, costs, and other relevant considerations;
7. *Optimization of the design and planning process*—the selection of an optimized, resource-effective sampling and analysis plan to meet established performance criteria; here, the team reviews existing data and gathers additional information that is useful in further optimization and identification of gaps that may be present as a result of the initial outputs.

The importance of establishing a sound sampling and analysis plan should be obvious as this is the foundation for the implementation and assessment phases. The quality of the data generated can then be assessed using the data quality indicators discussed in Chapter 2.

3.2.2 Traceability

A key aspect of QA is traceability—the ability to trace the history, application, or location of an entity by means of recorded identifications. Think of traceability as a paper trail in a sense. More specifically, we can break this down further as shown in the following two examples:

1. *Calibration*—traceability relating, for example, instrumentation to national or international standards, primary standards, properties, or reference materials;
2. *Data collection and analysis*—traceability relating collected results and calculations back to specified requirements for the quality of the investigation.

As discussed, traceability is particularly important with respect to standards that are used to calibrate an analytical instrument, where the accuracy of the results is vital. This is achieved by, for example, incorporation of a certified reference material

(Section 3.8) with a well-defined traceability linkage from an accredited entity. In addition, it is an important aspect of the data collection and analysis process, and a key consideration of sample handling and custody.

3.2.3 Sample Handling and Chain of Custody

Sample handling and chain of custody (COC) procedures are necessary to establish the traceability of environmental samples from the time of collection, through storage, and up to the time of analysis. Popek (2003) eloquently details the seven steps of a sample's life:

1. The sample is planned or "conceived."
2. The sampling point(s) is/are identified.
3. The sample is collected or "born."
4. The sample is transferred to the laboratory.
5. The sample is analyzed.
6. The sample expires and is discarded.
7. The sample is reincarnated as chemical data.

A COC document is necessary for proper documentation of the possession of samples during the seven-step process. Identification is paramount and every attempt must be made to label each sample properly. Samples that are removed from the sampling location and transported to a laboratory must be identified by a sample tag or label. Each person involved in the chain of possession must sign a COC form whenever sample custody is relinquished or accepted. How would investigators treat *in situ* type measurements? If *in situ* measurements are made, the data should be recorded in field notebooks or field data cards. Supporting information, including any field observations, weather, or other remarks, should be noted. The COC document serves a legal purpose and also helps to ensure the integrity of the samples of interest. Sample integrity can be fortified by following the generated sampling plan, allowing only trained investigators to perform the sampling, using appropriate sampling techniques, and preventing unwanted contamination or sample adulteration during storage. At a minimum, the COC document should contain the sample number, sample collector, location, date, and time of sample collection, sample matrix, signatures of investigators involved in the chain of possession, and dates of sample relinquishment. An example of a COC document is provided in Appendix IV.

> *In situ* literally means in the same position. In an environmental context, measurements are taken with a probe or analytical instrument directly in the matrix of interest.

3.2.4 Accreditation

Accreditation is a key aspect of quality assurance that formalizes the competence of a testing laboratory to conduct relevant tests. Within the framework of an accreditation procedure, a testing laboratory subjects itself to examination by an independent accredited authority. If the authority sees fit and awards accreditation, the laboratory can confirm its status as possessing the necessary expertise to conduct proper analyses, and guarantee compliance with the requisite standards, guidelines, and other statutory regulations. A representative selection of key regulatory bodies, many of which are accredited authorities, and their guidelines and standardization procedures are highlighted in Table 3.1.

The International Standardization Organization (ISO) is one of the most recognized and universally established authorities listed in Table 3.1. The ISO is a nongovernmental organization that has no legal authority to enforce the implementation of the many standards it sets. However, countries have the option to adopt ISO

TABLE 3.1

Key International Regulatory Bodies Governing Standardization and Accreditation of Laboratory Procedures

Body	Date Established	Services/Goals
Association of Official Analytical Chemists (AOAC)	1884	Multilaboratory and independent laboratory validation for nonproprietary and commercial proprietary methods.
Cooperation of International Traceability in Analytical Chemistry (CITAC)	1993	To help improve traceability of the results of chemical measurements for worldwide laboratories.
Environmental Protection Agency—United States (U.S. EPA)	1970	Development of laws, guidelines, and regulations to protect human health and the environment.
European Committee for Normalization (CEN)	1961	To contribute to the objectives of the EU for environmental protection, exploitation of research, and development programs.
Food and Drug Administration— United States (FDA)	1927	To ensure the safety of America's food supply, and to make safe, effective, and affordable medical products available to the public.
Food and Agricultural Organization–World Health (FAO/WHO)	1945	To aid developing, and other, countries improve agriculture, forestry, and fisheries practices, and to ensure proper nutrition for all.
International Laboratory Accreditation Cooperation (ILAC)	1977	To develop international cooperation for facilitating trade by promotion of the acceptance of accredited test and calibration results.
International Standardization Organization (ISO)	1947	World's largest developer of international standards.
International Union of Pure and Applied Chemistry (IUPAC)	1919	Method validation in a wide variety of chemically-related fields.
Organization for Economic Co-operation and Development (OECD)	1961	One of the world's largest and most reliable sources of comparable statistics, and economic and social data.

standards, especially those concerned with health, safety, and the environment. A published standard concerning the requirements for the competence necessary for testing and calibration (including sampling) in a laboratory setting is ISO 17025. Encompassed within this standard are 15 management and 10 technical requirements outlining the necessary steps that a laboratory should take with regard to the development of their management systems for quality, administrative, and technical operations.

3.2.5 GOOD LABORATORY PRACTICE

Another key organization listed in Table 3.1 is the Organization for Economic Co-operation and Development (OECD). They are involved in a variety of social and economic activities including the development of the *good laboratory practice* (GLP) assurance system—a framework within which laboratory studies are planned, performed, monitored, recorded, reported, and archived. These studies are undertaken to generate testing procedures for the mutual acceptance of data (MAD) by which the hazards and risks to users, consumers, and third parties, including the environment, can be assessed for industries and applications (Koëter 2003). GLP aids in assuring regulatory authorities that the data generated and submitted are a true reflection of the results obtained during a particular study, and can therefore be utilized in making risk and safety assessments. GLP principles and the abbreviated scope of each include (OECD 1986):

1. *Test facility organization and personnel*—The management ensures that the principles of GLP are complied with in the test facility and that appropriate standard operating procedures (SOPs—see principle 7) are established and followed.
2. *Quality assurance program*—The test facility sets in place a quality assurance program to ensure that studies are performed in compliance with the principles of GLP.
3. *Facilities*—Test facilities should be of suitable size, construction, and location to meet study requirements. A sufficient amount of separation from test materials and waste disposal is needed to prevent contamination and mix-ups.
4. *Apparatus, material and reagents*—Apparatus should be of appropriate design, adequate capacity, and periodically cleaned, maintained, and calibrated. All reagents should be labeled to indicate source, identity, concentration, preparation date, earliest expiration date, and storage instructions.
5. *Test systems*—Apparatus for the generation of chemical/physical data should be of appropriate design and capacity. Reference substances should be utilized to ensure integrity. For biologicals, proper conditions should be maintained and be in line with national regulatory requirements for import, collection, care and use.
6. *Test and reference substances*—Proper receipt, handling, sampling, and storage procedures should be maintained. Each test and reference substance should be identified and appropriately labeled. A sample for analytical purposes from each batch of test substances should be retained for studies lasting longer than four weeks.

7. *Standard operating procedures (SOPs)*—Test facilities should have SOPs to ensure the quality and integrity of the data generated in the course of the study. SOPs should be available for, but not limited to, the following laboratory activities: test and reference substances, apparatus and reagents, record-keeping, reporting, storage and retrieval, test systems, QA procedures, and health and safety precautions.

8. *Study performance*—Each study should include a study plan that exists in a written format prior to initiation of the study. The study plan should contain, but not be limited to, the following information: identification of the study, test and reference substances, sponsor and test facility information, dates, test methods, relevant issues (e.g., justification, dose levels, experimental design) and records.

9. *Reporting*—A final report should be prepared for the study, with the study director signing and dating this report. All corrections or additions to the report should be in the form of an amendment.

10. *Storage and retention of records and material*—Archives should be designed and equipped for the accommodation and the secure storage of raw data, study plans, final reports, samples, and specimens. All material should be indexed in a logical manner. All material should be retained according to a period specified by the appropriate authorities.

The OECD GLP guidelines are considered the international standard, with the European Community (EC) readily adopting the OECD principles into their laws and regulations for all the nonclinical safety studies that are listed in sectoral directives. The U.S. EPA has established a good laboratory practice standards (GLPS) compliance-monitoring program to ensure the quality of the test data submitted to the agency in support of the Federal Insecticide, Fungicide and Rodenticide Act (FIFRA), and with regard to certain testing agreements under the Toxic Substances Control Act (TSCA).

3.3   ENVIRONMENTAL SAMPLING PROTOCOL—GENERAL CONSIDERATIONS

In Chapter 2, we briefly discussed the concept of environmental sampling and how Pierre Gy had provided key evidence as to why investigators need to be critically aware of the sampling process they design and also eloquently describing the errors associated with the complete environmental analysis process. It is therefore essential to develop a sampling protocol that considers the identification of scientific objectives, safety issues, and budget constraints. Note that it is beyond the scope of this book to cover all aspects of the sampling processes in the myriad of complex environmental matrices encountered. Whole books, or major portions of others, provide such detail, and reading is recommended for background purposes (Keith 1991, Gy 1998, Popek 2003, Zhang 2007). Regardless of the matrix, an essential requirement of any sampling protocol is for the sample to be representative of the matrix from which it originates. It is therefore essential to adopt a well-organized protocol that retains, as closely as possible, the original composition of the matrix of interest. The protocol should be kept as simple as possible while minimizing the possibility of contamination or interference.

3.4 QUALITY ASSURANCE: SAMPLE COLLECTION, PREPARATION, AND STORAGE

It has been traditionally assumed among environmental scientists that the quality of data and the associated errors, in environmental analysis, are primarily attributed to the nature of the analytical methods. Such an assumption leads to an underestimation of the importance of other factors including:

1. sampling design and collection;
2. sample preparation (e.g., subsampling, aliquotation);
3. cleaning protocols;
4. sample pretreatment and preservation (e.g., filtration, digestion, centrifugation);
5. transport and storage.

Wagner (1995) presents a comprehensive review of the approaches and methods for quality assurance and quality control in sample collection and storage for environmental field investigations, and carefully examines the items listed here. Let us consider a few of them in more detail.

3.4.1 SAMPLING DESIGN

Multiple sampling techniques are utilized in environmental analysis depending on site-specific conditions, analytical technique used for analysis, and program objectives. Rather than rely on haphazard sampling methods that can lead to biased results, investigators incorporate probability sampling methods. Such methods allow for determination of spatially distributed variables and sampling along temporal and spatial scales. Six common probability sampling methods and brief descriptions of each are provided:

1. *Simple random sampling*—one that gives each sample unit an equal probability of being selected.
2. *Stratified random sampling*—splitting the population into sections or strata and choosing a random sample from each; note that environmental samples are often stratified by land type, terrain, and zones of contamination.
3. *Systematic sampling*—in which periodicity of the sampling cycle corresponds to natural cycles or trends.
4. *Cluster sampling*—useful when population components are found in groups or clusters; sometimes considered cheaper and investigator-friendly when compared to simple random sampling.
5. *Multistage sampling*—when population units fall within a staged structure category.

6. *Double sampling*—which comes in three forms: ranked set sampling, weighted double sampling, and double sampling with ratio estimation. All three methods have been shown to increase precision and lower sampling costs when compared to simple random sampling.

7. *Composite sampling*—another technique whereby investigators take several physical samples from an area, mix them and take a subsample of the mixture. Note that this is physical averaging rather than mathematical averaging, and is not considered the best sampling technique when information on high values or variability is desired. Also, in the Gy sense, statistical random sampling cannot be carried out with bulk materials because individual units or elements cannot be selected one at a time and at random (Gy 1998).

EXAMPLE PROBLEM 3.1

1. Which of the following sampling techniques is most appropriate for sampling the diurnal fluctuations of nutrient concentrations in a stream?

a. Random sampling
b. Simple stratified sampling
c. Systematic sampling

Answer: c. Systematic sampling as it corresponds to natural cycles or trends. The amplitudes of nitrogen and phosphorus cycles in such systems have been shown to be affected by, for example, diurnal fluctuations in water temperature and/or incident radiation (Scholefield et al. 2005).

Smith (2004) provides a detailed look at audits and assessment of sampling on the basis of the Gy system, ultimately concentrating discussion on the importance of procedures, equipment, and practice. Here, the author describes three steps in the audit and assessment of sampling systems:

1. to review the sampling procedures for the complete process;
2. to observe how the samples are being taken and at what locations; to determine if this process follows the sampling procedures outlined previously; and to determine if the appropriate equipment is being implemented and implemented correctly;
3. to report, both orally and in writing, to the sponsor of the audit and other key personnel.

Overall, a well-designed sampling program is intended to ensure that the collected samples and resultant data are representative of the target population. The efficient use of the resources of time, money, and personnel to meet the ends of the study is of paramount importance. Safety and geographical constraints are concerns that should also be considered.

3.4.2 SAMPLE COLLECTION, PRESERVATION, AND STORAGE

The overall effectiveness of any sample preservation and storage protocol depends on an assortment of factors including the nature of the sample matrix, cleaning procedures for sample devices and containers, container material and size, temperature, chemical treatment (e.g., addition of chloroform), and physical treatment (e.g., filtration, irradiation of sample, and pasteurization) aspects (Zhang et al. 2002). This is especially true in aquatic systems where a variety of chemical, biological, and physical processes take place continually. In these systems, samples should be collected from the water column at a series of depths and at cross-sectional locations as individual grab samples or through the use of automated samplers for time-series acquisition (Worsfold et al. 2005). It is also vital to avoid boundary areas, for example, the confluence of streams or rivers and below sewage treatment works, unless their impact on the system is being investigated. Other water bodies pose additional complications and these must be considered. In lakes and reservoirs, for example, representative sampling is often difficult due to environmental heterogeneity, both spatial and temporal (e.g., thermal stratification). In order to study biogeochemical cycling in stratified water bodies, appropriate depth profiling is required.

Location and frequency must also be considered when designing a sampling protocol. Site selection will ultimately depend on the problem to be addressed, and safety and accessibility are of paramount importance. The frequency of sampling (e.g., seasonal, continual) will depend on the scientific objectives and is often constrained by cost. For example, the highest phosphorus loadings in rivers and streams are generally correlated with intense, short-term discharges during the autumn and winter months, while the lowest loadings occur in the summer months when discharge is low and biological activity is high (Hanrahan 2001). In-water processes that affect phosphorus concentrations, which must also be considered, include plant, algal, and bacterial turnover, anthropogenic inputs (e.g., sewage effluent), matrix considerations (e.g., water hardness), and resuspension of bottom sediments from increasing river discharge (Worsfold et al. 2005).

Prior to any sampling campaign, it is essential to adopt an efficient cleaning protocol for all sampling equipment and storage bottles, and to continue this throughout the study. The walls of sample containers, for example, are excellent substrates for bacterial growth and therefore rigorous cleaning of all laboratoryware is necessary. In addition, sampling blanks should be taken to monitor and control the sampling process.

Preliminary treatment often involves filtration, which differentiates between the dissolved phase (operationally defined as that fraction which passes through a 0.45 or $0.2\,\mu m$ filter) and suspended matter collected on the filter (Horowitz et al. 1992). Filtration with a $0.2\,\mu m$ filter is preferred as it removes the majority of bacteria and plankton that would otherwise alter species concentrations during storage (Horowitz et al. 1992). As with sample containers, the filtration apparatus (including individual filters) must be cleaned prior to use with a similar acid wash/ultrapure water rinse procedure. The filtration procedure can be carried out under positive pressure or vacuum. However, excess pressure gradients should be avoided as rupture of algal cells and the subsequent release of intracellular contents into the sample could then occur (Worsfold et al. 2005). The choice of a sample preservation technique also varies widely and depends primarily on the analyte to be measured. Preservation methods include, but are not limited to, physical (i.e., refrigeration, freezing, and deep-freezing) and chemical (i.e., addition of chloroform, mercuric chloride, and acidification) techniques.

There are a number of ways in which specific matrix characteristics (e.g., hardness, salinity, dissolved organic matter, and bacterial nutrient uptake) have the potential to affect the integrity of collected samples. With this in mind, Gardolinski et al. (2001) examined the long-term (247 days) trends in measured total oxidized nitrogen (TON) and filterable reactive phosphorus (FRP) concentrations in natural waters with contrasting matrix compositions—salinities of 0.5% (River Frome, UK), 10% (Tamar Estuary, UK), and 34% (Plymouth Sound, UK). The aim of this study was to apply recommended storage protocols systematically ($-80°C$, $-20°C$, $4°C$ and $4°C$ and $-20°C$ with 0.1% [v/v] chloroform) for up to 247 days prior to analysis. A representative list of storage protocols for nutrients in natural waters is presented in Table 3.2.

Samples were collected and filtered utilizing many of the techniques and QA measures mentioned in Sections 3.4.1 and 3.4.2. Concentrations were determined using a segmented flow analyzer. Results of this study show that it is difficult to select one reliable treatment for all sites due to the variability in physico-chemical parameters. Ultrapure-water *blanks* and *controls* (see Section 3.5) were utilized to monitor the effects of storage treatments on artificial, salinity-matched matrices. For the controls (Figure 3.1), dashed lines represent $\pm 3s$ of the mean of each of three measurements of the two replicates ($n = 6$) analyzed on day 0. The error bars for all control data points represent $\pm 2s$ of the mean of three replicate measurements ($n = 3$). For TON, all controls remained stable for the duration of the experiment (247 days) with the exception of the 35% control stored at $4°C$, which showed a decrease (loss of 23.1 μM) between days 84 and 247. This occurrence is likely to be the result of nutrient uptake by the bacteria present in the seawater used in the preparation of the controls. For FRP, all controls remained stable when stored in the freezer but there was a marked deterioration (loss of 0.76 μM) when stored at $4°C$.

For the actual samples, let us concentrate on FRP in all matrices studied. Figure 3.2 shows the measured FRP concentration over the course of the 247-day study under all chosen storage conditions. Dashed lines represent $\pm 3s$ of the mean of each measurement of three replicates of the relevant samples ($n = 3$) analyzed on day 0. The error bars for all sample data points represent $\pm 2s$ of the mean of six replicate measurements ($n = 6$). For the most part, FRP concentrations in all subsamples behaved similarly when comparing sites and treatments. A closer examination of the River Frome site (0.5%) revealed that samples remained stable for the duration of the experiments when stored at $4°C$ with chloroform. For the $4°C$ storage without chloroform, the day 0 concentration (2.88 ± 0.43 μM) was maintained until day 28 after which there was a steady reduction to 0.005 ± 0.01 μM, that is, almost complete removal. This can be linked to bacterial uptake of FRP, as was observed in the controls (Figure 3.1). Results from the two freezer treatments showed a noticeable decrease (day 1) after the freezing/thawing process, which can be explained by physicochemical alteration of the samples. This was likely due to the coprecipitation of

TABLE 3.2

Representative Protocols for the Storage of Dissolved Nutrients in Natural Waters

Nutrient Species[a]	Matrix	Method of Storage	Experimental Details	Reference
FRP	Distilled, tap and lake water	Refrigerator (4°C)	Polypropylene and polycarbonate containers were appropriate for storage. Glass containers sorbed phosphorus within 1–6 hours.	Ryden et al. (1972)
FRP, TP	Open ocean water	Frozen, with/without $HgCl_2$ (120 mg/L), phenol (4 mg/L), and acid (pH 5)	No significant changes in TP concentration when samples were frozen with/without acid.	Morse et al. (1982)
NH_4 and NO_3	Precipitation and lake water	Refrigerator (4°C)	Significant changes in concentration were observed after 1 day of storage.	Vesely (1990)
TP, TDP, FRP, and TRP	Lake water	Refrigerator (4°C)	No significant change in TP samples for up to six months.	Lambert et al. (1992)
FRP	Soil leachates	Room temperature (5–19°C), refrigeration (4°C), frozen (−20°C) with/without $HgCl_2$ (40–400 mg/L) and H_2SO_4	Changes occurred within two days for all samples with smallest changes in samples stored at room temperature or 4°C.	Haygarth et al. (1995)
FRP, TON	Riverine, estuarine and marine	−80°C, −20°C, 4°C and 4°C and −20°C with/without 0.1% (v/v) chloroform	4°C with chloroform proved to be reliable. Samples from chalk-based catchment should not be frozen.	Gardolinski et al. (2001)

[a] FRP = filterable reactive phosphorus; TP = total phosphorus; TDP = Total dissolved phosphorus; TRP = Total reactive phosphorus; TON = total oxidized nitrogen.

Source: Modified from Gardolinski et al., *Water Research,* 35, 3670–3678, 2001. With permission from Elsevier.

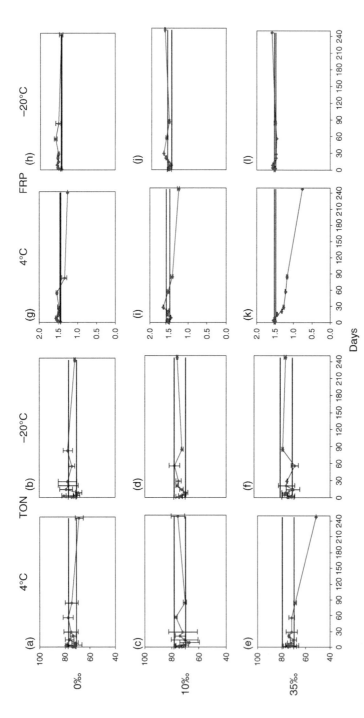

FIGURE 3.1 Concentrations (μM) of TON and FRP in control samples with varying salinities stored at 4°C and −20°C for up to 247 days. Error bars represent ±2s ($n = 6$). Dotted lines represent ±3s ($n = 9$) on the initial measurements (day 0) of the control samples. (From Gardolinski et al., *Water Research*, 35, 3670–3678, 2001. With permission from Elsevier.)

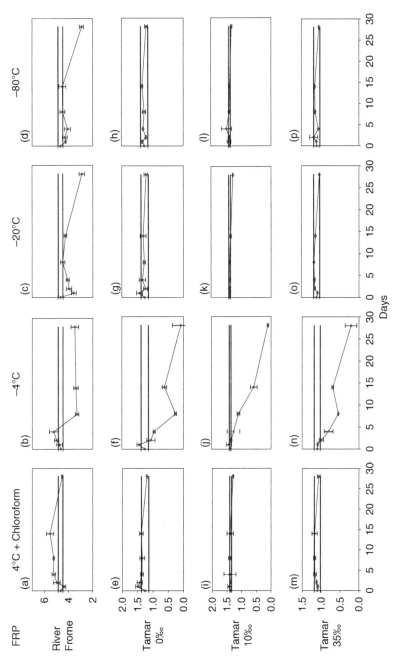

FIGURE 3.2 Concentrations (µM) of FRP in subsamples from all sites tested at 4°C (with and without chloroform), −20°C and −80°C for up to 28 days. Error bars represent ±2s (n = 6). Dotted lines represent ±3s (n = 9) on the initial measurements (day 0) of the samples. (From Gardolinski et al., *Water Research*, 35, 3670–3678, 2001. With permission from Elsevier.)

inorganic phosphate with calcite, thus removing FRP from the solution. Such events highlight the need to consider the chemical matrix as well as biological factors when developing storage protocols.

3.5   FIELD QA/QC SAMPLES

Previously, we discussed the potential for errors during the sample collection, preservation, and storage processes. Aspects of sampling equipment (analyte carryover), cross-contamination between samples, as well as sample representativeness must also be considered. To test for the absence or presence of these errors, and for accuracy and precision, the following QA/QC procedures are routinely utilized:

1. *Equipment (rinsate) blanks*—used to assess sample representativeness and detect sample cross-contamination. They are prepared in the field prior to sampling by implementing pre-cleaned equipment. Contaminants typically found include phthalates, methylene chloride, acetone, oily materials, PCBs and VOCs. Typically, one equipment blank is collected per type of sampling per day.
2. *Trip blanks*—used in VOC determinations. Unlike field blanks, they are prepared in volatile organic analysis (VOA) vials in the laboratory prior to field collection and never opened in the field. They are kept with the samples and sample containers to determine whether contaminants have been introduced while being handled in the field and during transport. Typically, one trip blank is utilized for each VOC method performed.
3. *Filter blank*—a blank solution (usually purified laboratory water) that is filtered in the same manner and through the same filter apparatus used for an environmental sample of interest.
4. *Field duplicates* (*soil and water*)—used to assess errors associated with sample heterogeneity, sampling and analysis precision, and sample representativeness. Note that this can increase overall sampling and analysis costs. Typically, one duplicate sample is utilized for every 10 field samples.
5. *Temperature blanks*—utilized to provide accurate measurements of sample temperature upon arrival at the laboratory. Typically, one temperature blank per cooler of samples is utilized.
6. *Background samples*—used to detect and quantify contaminant concentrations due to a contaminated source or site. Typically, one background sample is collected per different matrix examined.
7. *QA sample*—replicates of field samples analyzed by two different laboratories. The purpose is to establish data comparability. Typically, one sample is utilized for every 10 field samples.
8. *Split samples*—aliquots of the same sample(s) taken from the same sample container after mixing or compositing. Note that they are analyzed independently and used to test intra- or inter-laboratory precision.
9. *Blind sample*—sample(s) submitted for analysis where the composition or origin of the sample is known to the submitting investigator but not to the analyst.

10. *Matrix spikes*—Quality control samples employed to evaluate the effect of a particular sample matrix on the overall accuracy of a measurement. Recovery of the matrix spike provides such an indication and is calculated as follows:

$$\frac{A_{ms} - A_{fs}}{A_a} \times 100 \qquad (3.1)$$

where A_{ms} = amount of target analyte measured in the matrix spike sample, A_{fs} = amount of target analyte measured in the field sample, and A_a = amount of target analyte spiked into the matrix spike sample.

11. *Matrix spike duplicates*—split samples spiked with identical concentrations of target analyte(s) prior to sample preparation and analysis. They are used to determine the precision and bias of a method in the same sample matrix.

3.6   LABORATORY AND INSTRUMENTAL METHODS: QA/QC SAMPLES

We learned from Chapter 2 that the quality of data (accuracy and precision) must be established and known before it can be used for decision-making purposes. This ideal situation is created by a validated analytical system operating in a state of statistical control. To maintain this control, effectively designed and implemented QC procedures are paramount in accredited routine laboratories. An effective quality assured program includes many of the following elements:

1. *Calibration procedures*—as discussed in Chapter 2, calibrations are vital in standardizing an instrument's response prior to quantitative analysis. Note that a second source confirmation (single point standard of a second source) is often used as a QA to determine if there are any detectable errors in the initial calibration. All analytical instruments must be calibrated, or their calibration verified, before they are used to provide quantitative environmental results.

2. *Initial calibration blanks*—serve as a point in the initial calibration and is prepared with the same reagents used in the preparation of standards and samples minus the analyte of interest. They are used to give a null reading for the instrument response when performing a calibration curve, and to establish instrument background.

3. *Instrument blank*—the reagent used in the final preparation of the samples, which is injected into the analytical instrument roughly every 10 samples after higher concentrated standards and samples have been run. Such a procedure aids in detecting carryover in the system.

4. *Method blank*—analyte-free matrices (e.g., reagent water) that are carried out from initial sample preparation to the final analysis. Method blanks aid in determining whether or not laboratory contamination causes false positive results in relevant analyses.

5. *Continuing calibration blank*—the initial calibration blank that is reanalyzed throughout the entire analytical run to aid in assessing baseline drift potential.

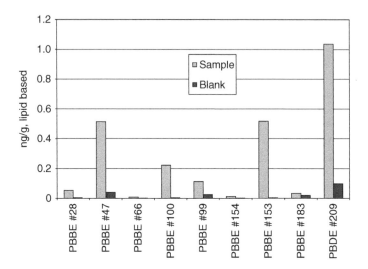

FIGURE 3.3   (**See color insert following page 206.**) Comparison of PBDE levels in blank and fish samples. (From Päpke, O., Fürst, P., and Herrmann, T., *Talanta*, 63, 1203–1211, 2004. With permission from Elsevier.)

Consider an application by Päpke et al. (2004) who utilized the analysis of laboratory blanks as a QC procedure for the determination of polybrominated diphenylethers (PBDEs) in biological tissues. PBDEs are structurally similar to dioxins, polychlorinated biphenyls (PCBs), and polybrominated biphenyls (PBBs) and are becoming increasingly widespread in the environment. The analysis of blank samples in such studies is vital, considering the potential for the contamination of solvents and adsorbents during PBDE analysis. Figure 3.3 shows blank samples compared with slightly contaminated fish samples. Overall, results indicate a relatively low influence of the blank on the sample values. The repeatability of the laboratory blank samples calculated for a typical sample weight of 0.15 g of lipid is shown in Table 3.3.

6. *Continuing calibration verification (CCV) standard*—primary calibration standard(s) that are reanalyzed with samples of interest to verify the continued calibration of the analytical system. These standards are typically run at the beginning and end of an analytical run, and often after every 10 to 15 samples in large experimental runs.

7. *Laboratory control samples*—a general class of samples that includes blank spikes and laboratory fortified blanks. Laboratory control samples (LCS) are prepared with analyte-free matrices fortified (spiked) with known concentrations of target analytes and carried through the entire sample preparation and analysis procedures. Such samples are typically run for every preparation batch, for up to 20 field samples, and aid the evaluation of analytical accuracy and the determination of laboratory precision between the LCS and LCS duplicates.

8. *Laboratory duplicates*—as with field duplicates, these are two aliquots taken from the same sample container that are processed and analyzed separately.

TABLE 3.3
Repeatability of Laboratory Blank Samples

PBDE #	Mean (ng g^{-1} Lipid Based)	n	Standard Deviation (ng g^{-1} Lipid Based)	R.S.D. (%)
28	0.0095	13	0.0057	60
47	0.063	13	0.041	65
66	0.0032	3	0.0012	36
99	0.035	8	0.028	78
100	0.0067	12	0.0080	119
153	0.0086	12	0.0092	106
154	0.0051	4	0.0046	89
183	0.014	12	0.010	75
209	0.078	10	0.034	43

Source: Modified from Päpke, O., Fürst, P., and Herrmann, T., *Talanta*, 63, 1203–1211, 2004. With permission from Elsevier.

Results are used to determine analytical precision for a given matrix from sample preparation to final analysis.

9. *Instrument duplicates*—two aliquots taken from the same extract or digestate and analyzed in the laboratory in a duplicate fashion to measure instrument precision.
10. *Laboratory matrix spikes*—similar to field matrix spikes but prepared and utilized in the laboratory setting. Here, a predetermined stock solution of the analyte(s) is added prior to any sample extraction/digestion procedure for subsequent analysis.
11. *Laboratory surrogates*—organic compounds (not found in the test samples) spiked into all blanks, calibration standards, and samples prior to analysis by GC or GC-MS. Percent recoveries can be calculated for each surrogate and internal control limits subsequently established.
12. *Instrument checks*—a variety of solutions (e.g., tuning solution, interference check solutions) utilized in routine analysis using GC-MS and ICP-MS instrumental techniques.
13. *Preparation and analytical batches*—a preparation batch is composed of 1–20 environmental samples of the same matrix prepared and analyzed with the same process and personnel. An analytical batch is a group of samples (e.g., environmental test samples, extracts) that are analyzed using the same method and personnel with the addition of a defined set of QC samples.

3.7 STANDARD ADDITION AND INTERNAL STANDARD METHODS

Discussion has centered on the importance of the matrix in environmental analysis and how interferences likely play important roles in affecting the magnitude of the analytical signal, and ultimately, in the determination of relative analyte concentrations. An important tool to aid analyses in interference-ridden samples is the process of *standard addition*, an alternative calibration procedure in which a known amount of a standard solution of analyte is added to a portion of the sample.

The standard addition method is widely utilized in atomic absorption and emission spectrometry studies, and has also found application in various electrochemical analyses. Although many forms of standard addition are possible, one of the most common involves the addition of one or more increments of a standard solution to sample aliquots containing identical volumes, and diluting to a fixed volume before measurement. Instrumental measurements are then made of each solution and corrected for any blank response to yield a net instrument response.

First consider the relationship between the analytical signal (A) and analytical concentration (c_x). Proportionality exists as shown by

$$A_x = nc_x \tag{3.2}$$

From the preceding discussion, diluting both the unknown solution and the unknown with the added standard to the same final volume (V_F) is necessary in order for the matrix to remain constant. Initially, $A_{x,F}$ is measured for the unknown sample diluted to V_F:

$$A_{x,F} = n\frac{c_x V_x}{V_F} \tag{3.3}$$

where V_x is the volume of the unknown solution of concentration c_x. Subsequently, a known amount of standard analyte is added to the unknown solution and diluted to V_F, with the resulting analytical signal measured by

$$A_{x+s,F} = n\left(\frac{c_x V_x + c_s V_s}{V_F}\right) \tag{3.4}$$

where V_s is the volume of the standard with a concentration c_s and $A_{x+s,F}$ is the analytical signal for the unknown plus the standard diluted to a final volume of V_F. Due to two unknowns having two separate equations, c_x is solved as

$$c_x = \frac{A_{x,F}}{A_{x+s,F} - A_{x,F}} \times \frac{V_s c_s}{V_x} \tag{3.5}$$

As usual, the signal is plotted and the errors assessed. A potential drawback of standard addition is the requirement of calibration graphs for each sample. This is in contrast to conventional calibration studies where one plot provides an assessment of multiple samples.

Silva and Nogueira (2008) utilized standard addition methodology for high performance liquid chromatography (HPLC) determination of triclosan in personal-care products and biological and environmental matrices. For sample assays, 25 mL of wastewater, toothpaste solution, and 1 mL of saliva were fortified with standard triclosan at desired concentration levels. Results showed that good overall calibration linearity (typically $r^2 = 0.999$) was achieved with no indication of observable matrix effects, and with a high degree of analytical selectivity, sensitivity, and accuracy.

Internal standards are in contrast to the standard addition method, as they require the use of a substance different from the measured analyte of interest. They are especially beneficial for chromatographic analyses in which the instrument response varies widely among runs, or if loss of analyte occurs, for example, during sample processing. In this method, a known amount of standard reference material is added to all samples, standards, and blanks with the intention of it assisting the identification and correction of the run-to-run discrepancies. This substance can then be used for calibration by plotting the ratio of the analyte signal to the internal standard signal as a function of the analyte concentration of the standards, ultimately taking into consideration a given dilution factor:

$$\frac{A_x}{[X]} = F\left(\frac{A_s}{[S]}\right) \tag{3.6}$$

where F is a response factor, and $[X]$ and $[S]$ are the concentrations of analyte and standard after mixing, respectively. See Example Problem 3.2 for a detailed explanation.

EXAMPLE PROBLEM 3.2

1. In a given experiment, a solution containing 0.050 M of analyte (X) and 0.044 M of standard (S) gave peak areas of 500 and 450 for A_x and A_s, respectively. Calculate the response factor, F.

Answer: Using Equation 3.6, F can be calculated as

$$\frac{A_x}{[X]} = F\left(\frac{A_s}{[S]}\right) \rightarrow \frac{500}{0.050} = F\left(\frac{450}{0.044}\right) \rightarrow F = 0.978$$

In order to determine the unknown concentration of X, 10.0 mL of 0.0190 M standard S was added to 10.0 mL of the unknown solution and diluted to 25.0 mL in a volumetric flask. Analysis of this mixture revealed peak areas of 550 and 590 for A_x and A_s, respectively. Calculate S as

$$[S] = (0.0190 \text{ M})\left(\frac{10.0}{25.0}\right) = 0.0076 \text{ M}$$

Incorporating F from above, X can now be calculated as follows:

$$\frac{A_x}{[X]} = F\left(\frac{A_s}{[S]}\right) \rightarrow \frac{550}{[X]} = 0.0076\left(\frac{590}{0.0076}\right) \rightarrow [X] = 0.0072 \text{ M}$$

Since X was diluted from 10.0 mL to 25.0 mL as the mixture with S was prepared, the original X in the unknown is calculated as

$$[X] = 0.0072 \text{ M}\left(\frac{25.0}{10.0}\right) = 0.018 \text{ M}$$

3.8   CERTIFIED REFERENCE MATERIALS

Certified reference materials (CRMs) are an important part of the method validation process and considered the best techniques for estimating the accuracy of a method. A CRM is a reference material of known concentration with similar matrix properties to the sample being analyzed and whose property values have been certified by a comprehensively validated procedure. CRMs have a use in:

1. verifying the accuracy of results obtained in a laboratory;
2. monitoring the performance of a method;
3. calibrating analytical methods;
4. the detection of errors in the application of standardized methods;
5. ensuring laboratory traceability.

   To serve as a measurement standard, a CRM must be stable. For this purpose, the material should undergo stability testing after it has been prepared (Van der Veen et al. 2001). Additionally, CRMs should also match with respect to the levels of concentrations of the analytes to be determined. Ideally, interlaboratory comparisons should be utilized to give an expression of uncertainty in measurements in the certification of reference materials.

   A review by Worsfold et al. (2005) highlighted a 2002 intercomparison study for the analysis of orthophosphate in MOOS-1, a natural seawater nutrient CRM produced by the National Research Council of Canada (NRCC) in direct response to the marine chemistry community (Clancy and Willie 2004). This analysis was carried out by 25 different laboratories participating in the NOAA/NRC 2nd intercomparison study for nutrients in seawater. Flow and manual methods of analyses were used, all based on the same spectrophotometric procedure. Eighteen of the 25 laboratories achieved satisfactory z-scores (see Chapter 2) for the determination of phosphate in seawater (Figure 3.4). Here, z-scores were calculated from the mean orthophosphate concentration, with the assigned value set at $1.6 \pm 0.21\ \mu M$. Overall, $|z| \leq 2$ represents

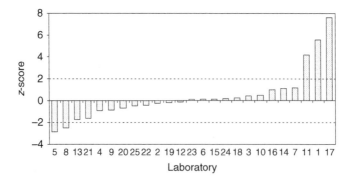

FIGURE 3.4   Plot of z-scores obtained by the NOAA/NRC 2002 intercomparison study for the analysis of orthophosphate in CRM MOOS-1. (From Worsfold et al., *Talanta*, 66, 273–293, 2005. With permission from Elsevier.)

TABLE 3.4

Example CRMS Available for the Determination of Phosphorus (P) Species in Environmental Matrices

CRM	Matrix	P Species	Concentration	Comments	Reference/Supplier
MOOS-1	Seawater	Orthophosphate	1.56 ± 0.07 µmol/L	Natural seawater CRM from Cape Breton Island, NS, Canada	Clancy and Willie 2004
QC RW1	Freshwater	Orthophosphate	100 µg/L	Artificial sample	Merry 1995
QC RW2	Freshwater	Total phosphorus	200 µg/L	Artificial sample	Merry 1995
Australian natural water CRM	Freshwater	Orthophosphate	27 ± 0.8 µg/L	Natural water sample obtained from Christmas Creek, Queensland, Australia	Queensland Health Scientific Services, Australia
BCR-616	Groundwater	Total dissolved P Orthophosphate	37 ± 1.2 µg/L 3.36 ± 0.13 mg/kg	Artificial groundwater sample	Institute for Reference Materials and Measurements, Belgium
BCR-684	River sediment	NaOH – extractable P HCl – extractable P Inorganic P Organic P	500 ± 21 mg/kg 536 ± 28 mg/kg 1113 ± 24 mg/kg 209 ± 9 mg/kg	Material collected from the River Po, Italy	Institute for Reference Materials and Measurements, Belgium
SRM® 1646a	Estuarine sediment	Total P	$0.027 \pm 0.0001\%$	Material dredged from Chesapeake Bay, USA	National Institute of Standards and Technology, USA

Source: Modified from Worsfold et al., *Talanta*, 66, 273–293, 2005. With permission from Elsevier.

the satisfactory *z*-score value for MOOS-1 (Clancy and Willie 2004). Examples of additional CRMs for the determination of phosphorus species in complex environmental matrices are provided in Table 3.4.

3.9   STATISTICAL QUALITY CONTROL CHARTS

Internal quality control (IQC) is the process of systemic monitoring of precision under repeated conditions to determine accuracy and random errors of quantitative investigations. A laboratory that does not maintain this control is one that cannot fully assess the accuracy and precision of its own analyses. IQC tools most often used are *control charts*—graphic analytic tools utilized by investigators to assess whether a process is in a state of statistical control. A variety of control charts are utilized and defined by user-identified purposes and by the type of data (variables and attributes):

1. mean, range, and standard deviation;
2. individual measurement and moving range charts (run chart, IR chart);
3. *p*-chart, *np*-chart, *c*-chart and *u*-chart;
4. uniformly weighted moving average (UWMA) chart and exponentially weighted moving average (EWMA) chart;
5. Cusum charts;
6. phase control, presummarized, multivariate, and Levey–Jennings charts.

   The concepts underlying such charts are that the natural variability in any process can be quantified within a set of control limits (upper and lower), and that variation exceeding these limits signals a special cause of variation. Control charts have three basic components including a centerline (typically the mathematical average of all samples plotted), upper and lower statistical control limits (typically 3σ), and performance data plotted over a given time frame. It is beyond the scope of this book to discuss every chart in detail. Specific applications, however, are discussed below.

   Suppose we are monitoring the performance of a spectrophotometric method over the course of 10 hours (one analysis per hour) by utilizing the orthophosphate CRM MOOS-1 ($1.56 \pm 0.07\,\mu M$) discussed here. Results of this analysis are presented in Figure 3.5a. Upper and lower control limits were based on $3\,\sigma$ from the center line (average of all 10 runs) with Zones (A, B, and C) added for further interpretation of special test cases. Test 1, for example, assesses one point beyond Zone A that helps detect shifts in the mean, an increase in the standard deviation, or a single aberration in the process. As shown, all individual analyses lie within $3\,\sigma$ of the center line, thus confirming the absence of any Type I or II errors. Figure 3.5b would be an example of a single aberration with the Test 1 data point labeled. The related moving range plot is shown in Figure 3.6. This chart displays moving ranges of two or more successive measurements. Moving ranges are computed for the number of consecutive measurements entered by the investigator. Typically the default range span is 2. Because moving ranges are correlated, these charts should be interpreted with care.

   Description and interpretation of complete special causes tests are found in Table 3.5. Investigators must pay close attention to the processes of *shift* and *drift* in any analytical

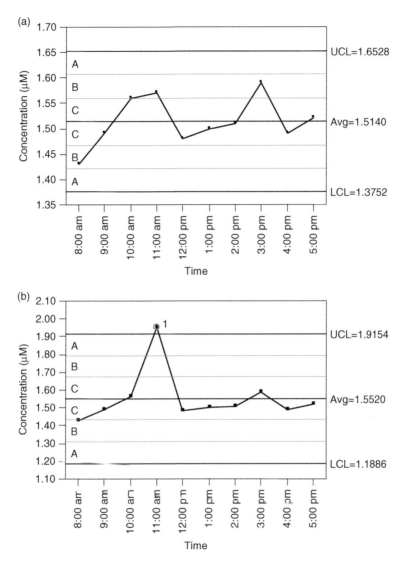

FIGURE 3.5 Statistical quality control charts for monitoring the performance of a spectro-photometric method. Results of this 10-hour study including upper and lower control limits are shown in (a). Panel (b) shows an example of a single aberration with the Test 1 data point labeled.

process. Shift may be the result of an introduction of new equipment, instability of the control sample utilized, use of a wrong standard, or wrong execution of the analytical method by the investigator. Often less conspicuous is the drift process, with an upward or downward trend, possibly indicating a drift in the mean or a gradual increase in the standard deviation, which may not be detected via F- and t-tests. Such drifts could be detected if control charts were implemented, especially if extended to larger numbers of

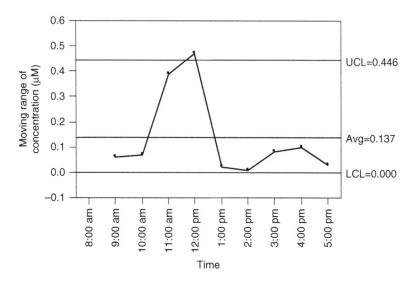

FIGURE 3.6   Moving average range plot from data generated in Figure 3.5.

TABLE 3.5

Description and Interpretation of Statistical Control Charts Special Case Tests

Test	Characteristics	Utilization
1	One point beyond Zone A.	To detect shifts in the mean, an increase in the standard deviation, or a single data aberration.
2	Nine points in a row in a single side of Zone C or beyond.	To detect a shift in the process mean.
3	Six steadily increasing or decreasing points.	To detect drift in the process mean.
4	Fourteen points in a row alternating up and down.	To detect systematic effects (e.g., alternating instruments or operators).
5	Two out of three points in a row in Zone A or beyond.	To detect shifts in the process average or increases in standard deviation.
6	Four out of five points in Zone B or beyond.	To detect a shift in the process mean.
7	Fifteen points in a row in Zone C (above and below center line).	To detect stratification of subgroups when the observations in a single subgroup come from various sources with different means.
8	Eight points in a row on both sides of the center line with none in Zone C.	To detect stratification of subgroups when the observations in a single subgroup come from various sources with different means.

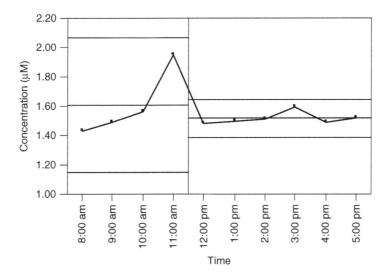

FIGURE 3.7 Phase control chart showing the aberrant data point (phase 1) and newly instituted CRM for further assessment (phase 2).

observations or measurements. Primary causes of drift include deterioration of reagents and equipment, and instability of the control sample. The aberrant data-point in Figure 3.6 could have been a result of the instability of the CRM. A new CRM must therefore be instituted, brought into the analytical process, and put through successive measurements. Such a scenario can be seen in the phase control chart depicted in Figure 3.7. Here, two phases are shown, one covering the aberrant data-point and the other, during successive measurements of the newly instituted CRM into the analytical process.

3.10 PROFICIENCY TESTING

Proficiency testing (PT) is the use of interlaboratory comparisons to effectively determine the performance of individual laboratories for specific tests of measurements. Here, comparisons of unknown samples are performed for testing instrument calibration, performance, operator-skill levels and consistency, and for use in accreditation. The latter, in general terms, refers to certification by a recognized body (e.g., ISO) of the facilities, capability, competence and integrity of a laboratory, service or operational group, or individual investigator to provide the specific service or operation needed. If the results of interlaboratory comparisons are in good statistical agreement, the obtained value is then likely to be a true approximation. Interlaboratory comparisons are useful for:

1. ascertaining the performance of an established or newly developed method;
2. measuring and improving the quality of a given laboratory (e.g., audits for accredited laboratories);
3. certifying the contents of a reference material.

During interlaboratory studies, individual laboratories compare their results with the consensus values generated by all the participating laboratories to assess the accuracy of its results, and hence, its analytical procedures. Criteria for acceptance in the form of tolerance ranges are also established. Different sample pretreatment methods and techniques of separation, for example, are compared. Consider an example application by Harner and Kucklick (2003), who initiated an intercomparison study to investigate consistency in reported concentrations of polychlorinated naphthalenes (PCNs). These compounds are a class of industrial chemicals similar to PCBs (Chapter 1, Section 1.3.3). Although no longer produced, and declining in the environment, certain congeners associated with combustion processes do exhibit an increasing presence (Meijer et al. 2001). The goal of this study was to check the quantification step of PCN analysis from nine laboratories in seven countries using varying methodologies, instrumentation, and types of standards employed. More specifically, the laboratories quantified individual homolog groups, Σ PCN (the sum of 2–8 chlorinated homologs) and selected congeners in two test solutions.

Table 3.6 lists the commonly used methods in addition to the quantification details for the nine laboratories participating in this study. Figures 3.8 and 3.9 show all the reported values for the sums of each of the homolog groups (Σ PCN) and individual PCN congeners for unknowns #1 and #2. Laboratory #8, who

TABLE 3.6

Instrument and Quantification Details for the PCN Intercomparison Study

Lab I.D. #	Analytical Instrument[a]	GC Column, Length (m)	Quantification External or Internal Standard	Quantification Single Response Factor (RF) or Calibration Curve (CC)	Type of Standard 1 = Mixture 2 = Individual Congeners 3 = 1 and 2
A	GCMS-EI	CP-SIL8CB, 50 m	Internal	CC	1
B1	GCMS-NICI	DB5-MS, 60 m	Internal	CC	1
B2	GCMS-NICI	DB5-MS, 60 m	Internal	CC	2
C	GCMS	DB5-MS, 30 m	Internal	CC	2
D	GCECD	DB5, 60 m	External	RF	2
E	GCMS	DB5, 30 m	Internal	RF	2
F	HRGC-HRMS	Ultra2, 25 m	External	RF	2
G	HRGC-HRMS	DB5-MS, 60 m	Internal	RF	3
H1	GCMS-NICI	DB5, 60 m	Internal	RF	1
H2	GCMS-NICI	Rt-βDEXcst, 30 m	Internal	RF	1
I	HRGC-HRMS	DB17, 30 m	External	RF	2

[a] GC = gas chromatography; MS = mass spectrometry; ECD = electron capture detection; HR = high resolution; EI = electron impact; NIC = negative ion chemical ionization.

Source: Modified from Harner, T. and Kucklick, J., *Chemosphere*, 51, 555–562, 2003. With permission from Elsevier.

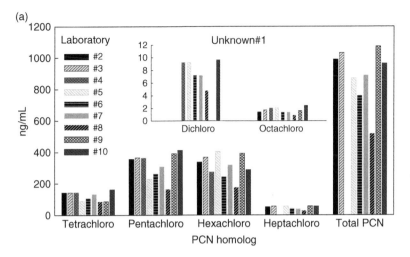

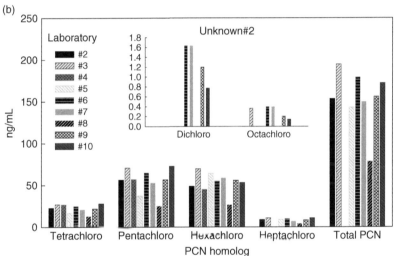

FIGURE 3.8 Concentrations for homolog group and total PCN in unknowns #1 (a) and #2 (b). (From Harner, T. and Kucklick, J., *Chemosphere*, 51, 555–562, 2003. With permission from Elsevier.)

performed the analysis by HRGC-HRMS, showed consistently lower values than the others for the PCN homologs, Σ PCNs and the PCN congeners for both unknowns.

Congeners are compound members of the same family.
Homologs are a group of structurally-related chemicals that have the same degree of chlorination.

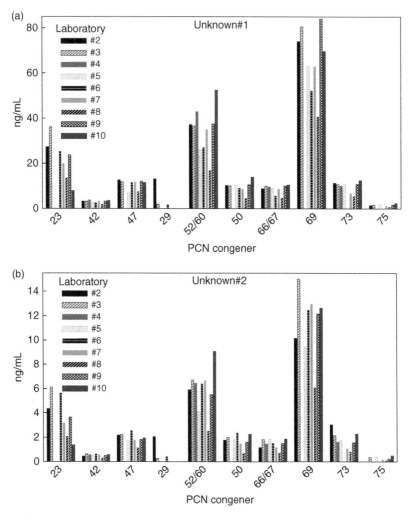

FIGURE 3.9 Concentrations for selected PCN congeners in unknowns #1 (a) and #2 (b). (From Harner, T. and Kucklick, J., *Chemosphere*, 51, 555–562, 2003. With permission from Elsevier.)

The overall variability among laboratories expressed as the percent standard deviation (SD) relative to the mean (RSD) was 20.2% and 22.7% for unknowns #1 and #2, respectively. The mean and standard deviations for all Σ PCN concentrations was reported as 883 ± 107 ng/mL and 153 ± 19 ng/mL for unknowns #1 and #2, respectively. These values were 80% and 83% of the expected values of 1104 ng/mL and 184 ng/mL based on the preparation of stock solutions. The participating laboratories, with one exception, quantified the Σ PCNs with consistency (11% RSD). Despite the positive initial findings, PCN congener determinations in the test solutions had higher variability (≈ 20–40%). Such results suggest the need for additional laboratory comparison studies that utilize CRMs.

EXAMPLE PROBLEM 3.3

1. Should interlaboratory comparison studies provide traceability?

Answer: Traceability is not always apparent in intercomparison studies, although it is a requirement in ISO/IEC Guide 43-1. The true value in most programs is generally not traceable to an independent entity. Note that the test materials should generally be similar in nature to those routinely tested by participating laboratories and widely available over a number of years. It is also important that results from intercomparison studies are cited in refereed sources (e.g. scientific journal articles) to enhance credibility of the data. See discussion on certified reference materials (Section 3.8).

3.11 DATA ARCHIVING, STORAGE, AND AUDITING

Environmental monitoring and research programs generate large amounts of information, ultimately providing valuable datasets in need of efficient management to help preserve its quality. More specifically, databases can be used as:

1. a means by which scientists and resource managers store and access environmental information;
2. an aid to developing environmental modeling software tools;
3. an aid to effective decision-making pertaining to environmental resource management and protection;
4. a way to provide valuable and technically sound data for the establishment and enforcement of legislative processes.

A variety of organizations and research institutions have developed environmental databases. For example, extensive databases have been generated from the National Marine Monitoring Programme (NMMP), the National Oceanic and Atmospheric Administration (NOAA), Marine Environmental Buoy Database [National Oceanographic Data Center (NODC)] Program, the EPA-led Great Lakes Environmental Database (GLENDA), and the Environmental Assessment and Review Process (EARP) developed by the Canadian Environmental Protection Agency (CEPA). Most databases are an integration of biological, chemical, and physical parameters that have been compiled using uniform methods with strict QA/QC practices applied. Review organizations such as the Committee on Data for Science and Technology (CODATA) are actively engaged in ensuring the quality, reliability, management, and accessibility of relevant data. There is therefore the potential to add to the repository of data already held by the relevant organizations providing incentives to adhere to QA/QC practices such as intercomparison studies in conjunction with the routine in-house use of, for example, CRMs.

Three types of audits are routinely employed: systems audits, performance audits, and data audits. Both in-house and external, independent auditors are routinely used for both field and laboratory operations. System audits are qualitative in nature and involve a walk-through of operations to ensure that all aspects of the QA program are followed. Performance audits also involve detailed walk-through and include quantitative checks, side-by-side collection and analysis of samples. Data audits involve several samples selected to be followed through the sample collection and analysis process. All documentation from sample collection to final analysis and storage are routinely checked.

3.12   MULTIVARIATE QUALITY ASSURANCE/CONTROL—INITIAL CONSIDERATIONS

It should now be obvious as to why investigators need to measure the quality of analytical processes and the multitude of environmental data generated. For example, how can *multivariate control techniques* be used to quantify simultaneously two environmentally relevant analytes showing strongly overlapped chromatographic peaks? Additionally, how can investigators discover hidden patterns and parameter relationships in large environmental datasets that visual inspection or traditional statistical methods cannot reveal? Categories of chemometric multivariate methods that are routinely incorporated in environmental analysis and are further discussed in this book include:

1. experimental design;
2. regression analysis;
3. exploratory analysis;
4. prediction and evaluation;
5. outlier detection and model validation;
6. classification.

Using multivariate chemometric methods in environmental analyses offers a faster and more precise assessment of the chemical and physical properties, the development of robust predictive models, and a true assessment of analytical measurement performance. Such methods are fully explored in subsequent chapters.

3.13   EXPANDED RESEARCH APPLICATION II—MONTE CARLO SIMULATION FOR ESTIMATING UNCERTAINTY INTERVALS FOR THE DETERMINATION OF NITRATE IN DRINKING WATER

We learned that certified reference materials and interlaboratory comparison studies can be used to control bias in the routine quality of analytical methods. This is especially true when considering the analytical methods and results of the examination of the content and regulation of the quality of water for human consumption. However, to truly assess uncertainty in the estimates of bias and precision, and to evaluate their confidence, more robust methods are needed. Escuder-Gilabert et al. (2007) examined the use of a Monte Carlo simulation tool to determine the quality of accuracy validation for a method to determine nitrate in drinking water.

3.13.1 STATEMENT OF PROBLEM

Laboratories performing only a single method-validation experiment are limited, as this merely provides information on bias and precision estimates (e.g., relative error, RSD), but not their variability if more validation experiments are performed.

3.13.2 RESEARCH OBJECTIVES

Considering the limitations of single method-validation experiments, the investigators presented an alternative approach, one evaluating the quality of bias and precision estimates as part of the overall validation process. Their objective was to perform several independent validations (after generating a data matrix) using the Monte Carlo simulation to gain information with regard to the uncertainty associated with the estimates of bias and intermediate precision of a method for determining nitrate in drinking water from a CRM.

3.13.3 EXPERIMENTAL METHODS AND CALCULATIONS

Note that all symbols for scalars, vectors, and matrices used throughout this study are provided in Appendix II. The basic theory behind data matrices and associated concepts are provided in Chapter 4 with matrix calculations reviewed in Appendix III. For Monte Carlo simulations, a computer application was programmed simulating the accepted true values for the reference material with the method characteristics for a given nitrate concentration level represented by: μ_0, E_0, $RSDr_0$, and $RSDrun_0$. From these, $RSDi_0$ is calculated as

$$RSDi_0 = (RSDr_0^2 + RSDrun_0^2)^{0.5} \tag{3.7}$$

The investigators used the zero in the subscript to differentiate the true values from the estimated values. Careful review of Appendix III will help understand the matrix calculations performed in this study. Here, a model initially proposed by Kuttatharmmakul et al. (2000) was used to generate the $\mathbf{X}_{N_t \times N_s}$ dataset, assuming normally distributed validation results consistent with the accepted true values. From this dataset, the error in percentage ($E = 100(\mu - \mu_0)/\mu_0$) and the relative standard deviation in intermediate precision [$RSD_i = (RSDr^2 + RSDrun^2)^{0.5}$] were calculated by adapting Equation 3.7. Finally, a number of simulations (e.g., 10^4 different validations) were defined by the investigators to obtain uncertainty estimates related to E and RSD_i. Studies by Herrador et al. (2005) provided the expanded uncertainly levels ($\pm U$; $CL \approx 95\%$) for this study. Uncertainty levels of 2.5% and 97.5% percentiles were used for $RSD_i \pm U(RSD_i)$. Intervals of 5% and 95% percentiles were used for $E \pm U(E)$.

A close look at the nitrate determination methods used in this study reveals analyses performed under ISO accredited (ISO 17025) and certified (ISO 14011) conditions. Haemodialysis water samples were spiked with 50 mg/L (μ_0) of nitrate (from a certified solution) to obtain experimental results under intermediate precision conditions. Here, duplicate analysis was performed for 12 different batches on three days by four different analysts.

Spiked samples are those fortified in the laboratory with a known concentration of analyte.

Can the design matrix be predicted from these sets of experiments? A quick glance at Chapter 4 would help in this endeavor. To aid understanding, let $X_{Nr \times Ns}$ represent a matrix with replicate $\times$ runs. In this study, a resultant $X_{2 \times 12}$ data matrix is thus formed. Analysis of the reference material was performed with the use of a spectrophotometric method adapted from APHA (2005), hence it became an internal standard operation procedure (SOP) that required validation before use as stated in ISO 17025 (ISO/IEC 2005). The data matrix generated from the $X_{2 \times 12}$ study is shown in Table 3.7.

3.13.4 RESULTS AND INTERPRETATION

The investigators first presented an overview of the Monte Carlo approach and generated experiments based on simulated laboratory information with the following criteria: $E_0 = 0\%$, $RSDr_0 = 5\%$, and $RSDrun_0 = 10\%$. Utilizing Equation 3.7 would then result in a $RSDi_0 = 11.18\%$, a value that the investigators believed would provide a significant run effect. They also assumed legislative bias and intermediate precision limit requirements of $E_{LIM} = \pm10\%$ and $RSDi_{LIM} = 15\%$, respectively. Figure 3.10 shows the simulated $RSDi$ vs. E results for $\mu_0 = 10$ (arbitrary units) at the specified conditions ($E_0 = 0\%$, $RSDr_0 = 5\%$ and $RSDrun_0 = 10\%$) for single experimental validations. Here, 10 independent simulation results are shown for Nr (2, 3, 4, 5, and 6) and Ns (4, 8, 12, 16, and 20). The selected Ns values would enable

---

TABLE 3.7

Experimental Results of the $X_{2 \times 12}$ Design Matrix

Batch	Analyst	Day	Time	Replicate 1 (mg/L)	Replicate 2 (mg/L)	Mean (mg/L)	Difference
1	1	1	11:00	52.18	51.83	52.01	−0.35
2	2	1	13:42	51.45	52.67	52.06	1.22
3	3	1	17:20	52.40	52.16	52.28	−0.24
4	4	1	18:00	51.79	51.50	51.65	−0.29
5	4	2	08:30	55.22	55.26	55.24	0.04
6	2	2	11:46	48.35	53.05	50.70	4.7
7	1	2	12:23	52.28	52.07	52.18	−0.21
8	2	2	13:28	48.57	50.63	49.60	2.06
9	4	2	17:45	51.38	49.87	50.63	−1.51
10	1	3	08:30	53.83	53.85	53.84	0.02
11	1	3	09:30	52.12	52.90	52.52	0.77
12	4	3	12:30	50.15	50.42	50.29	0.27

Source: Adapted from Escuder-Gilabert et al., *Analytical and Bioanalytical Chemistry*, 387, 619–625, 2007. With permission from Springer.

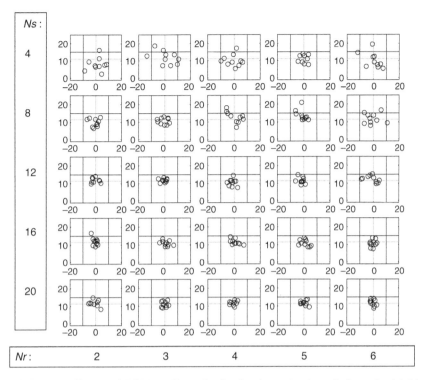

FIGURE 3.10 Simulated *RSDi* vs. *E* results for the determination of nitrate in drinking water. (From Escuder-Gilabert et al., *Analytical and Bioanalytical Chemistry*, 387, 619–625, 2007. With permission from Springer).

the development of experimental designs (Chapter 4) to determine the effects of investigator, equipment, and time. This simulation, however, had mixed results with some validation results that agreed with the criteria and others that did not. To alleviate this, the investigators recommended repeating the validation experiments numerous times (e.g., 10^4), ultimately allowing for proper computation of $E \pm U(E)$ and $RSD_i \pm U(RSD_i)$ uncertainty levels.

The investigators then performed actual experiments for validating the bias and intermediate precision of the nitrate method used for drinking water control. In this study, limits were fixed at $E_{LIM} = 10\%$ and $RSDi_{LIM} = 14\%$ (by adapting Equation 3.7) for graphical accuracy assessment. Using the results from Table 3.7, the investigators estimated $E = 3.83\%$. ANOVA was used to estimate precision and resulted in $RSDr = 2.19\%$ and $RSDrun = 2.56\%$. Assessment by a graphical uncertainty interval approach (Figure 3.11) revealed that both $E \pm U(E)$ and $RSD_i \pm U(RSD_i)$ intervals were within the 98/83/EC Directive limits.

3.13.5 SUMMARY AND SIGNIFICANCE

A novel, diagnostic graphical tool based on the Monte Carlo simulation has been developed by Escuder-Gilabert et al. (2007) and effectively applied to compare

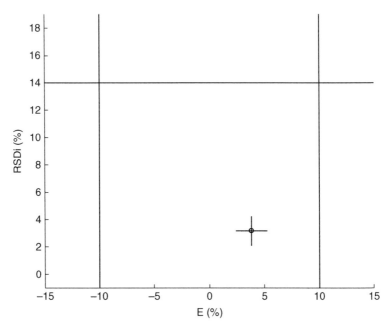

FIGURE 3.11 *RSDi* ± *U(RSDi)* vs. *E* ± *U(E)* uncertainty intervals after 104 simulations. Acceptation criteria limits are shown as solid lines (From Escuder-Gilabert et al., *Analytical and Bioanalytical Chemistry*, 387, 619–625, 2007. With permission from Springer).

uncertainty intervals for bias and intermediate precision in the analysis of nitrate in drinking water. The medium cost $\mathbf{X}_{2\times12}$ design chosen by the investigators was performed at a single concentration level (50 mg/L—the limit of nitrate in drinking water stipulated by the 98/83/EC Directive), and has proved to be extremely effective in assuring quality results.

3.14 CHAPTER SUMMARY

The highest quality environmental data are those that guarantee the requisites dictated by individual investigators, managers, policy-makers, and environmental health professionals. Such data, whether generated by laboratory analytical systems or field-based collection and analysis, are characterized by an incessant variability arising from the nature of complex matrices, laborious sample preparation and storage schemes, and modern, high-throughput instrumentation. Sound quality assurance programs incorporating both sampling and analysis must therefore be defined and implemented as part of the environmental monitoring process. Subsequently assessed quality assurance and quality control based on traditional statistical concepts are commonly employed, but do not provide full data analysis capabilities. Incorporating multivariate chemometric methods into environmental analyses offers a better assessment of analytical processes and a more efficient interpretation of environmental datasets in order to gain information about measurements, and to establish the neces-sary limits that must be accounted for.

3.15 END OF CHAPTER PROBLEMS

1. How do the concepts of quality assurance and quality control differ in their approaches?
2. What is the purpose of data transformation techniques in the context of environmental analysis and give some of the advantages? How does the choice of transformation method affect the analytic output?
3. What two statistical measures must be known in order to have confidence in a measurement result?
4. Discuss why there is no such thing as a standard protocol for storage of environmental samples for subsequent laboratory analysis.
5. A given environmental regulation requires random sampling for verification and quality control testing. Would a stratified random sampling approach meet the overall requirement of the regulation?
6. Consider the following control chart, which analyzes the performance of an IQC standard over the course of an experiment.
   a. What do the lines labeled UCL and LCL represent?
   b. Is this process under good statistical control?

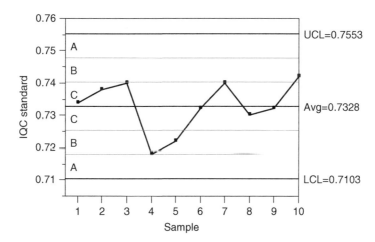

7. An investigator employed a single standard addition method to determine the concentration of analyte X in an unknown sample. Without standard addition, the absorbance measured 0.230. To 10.0 mL of sample, 0.50 mL of a 100 mg/L standard was added and properly mixed. The new absorbance measured was 0.390. What was the analyte concentration in the original sample?
8. What is the purpose of an interlaboratory comparison study? Explain how such a process is conducted.
9. Name the advantages of using multivariate chemometric techniques in quality assurance and quality control processes.

REFERENCES

American Public Health Association (APHA). 2005. Standard Methods for the Examination of Water and Wastewater, 21st Edition.

Clancy, V. and Willie, S. 2004. Preparation and certification of a reference material for the determination of nutrients in seawater. *Analytical and Bioanalytical Chemistry* 378: 1239–1242.

Escuder-Gilabert, L., Bonet-Domingo, E., Medina-Hernández, M. J., and Sagrado, S. 2007. A diagnostic tool for determining the quality of accuracy validation. Assessing the method for determination of nitrate in drinking water. *Analytical and Bioanalytical Chemistry* 387: 619–625.

Gardolinski, P. C. F. C., Hanrahan, G., Achterberg, E. P., Gledhill, M. C., Tappin, A. D., House, W. A., and Worsfold, P. J. 2001. Comparison of sample storage protocols for the determination of nutrients in natural waters. *Water Research* 35: 3670–3678.

Gy, P. 1998. *Sampling for Analytical Purposes.* Wiley: West Sussex.

Hanrahan, G., Gledhill, M., House, W. A. and Worsfold, P. J. 2001. Phosphorus loading in the Frome Catchment, UK: seasonal refinement of the coefficient modelling approach. *Journal of Environmental Quality* 30: 1738–1746.

Harner, T. and Kucklick, J. 2003. Interlaboratory study for the polychlorinated naphthalenes (PCNs): phase 1 results. *Chemosphere* 51: 555–562.

Haygarth, P. M., Ashby, C. D., and Jarvis, S. C. 1995. Short-term changes in the molybdate reactive phosphorus of stored soil waters. *Journal of Environmental Quality* 24: 1133–1140.

Herrador, M. A., Asuero, A. G., and González, A. G. 2005. Estimation of the uncertainty of indirect measurements from the propagation of distributions by using the Monte Carlo method: An overview. *Chemometrics and Intelligent Laboratory Systems* 79: 115–122.

Horowitz, A. J., Elrick, K. A., and Colberg, M. R. 1992. The effect of membrane filtration artifacts on dissolved trace element concentrations. *Water Research* 26: 753–763.

ISO/IEC 17025:2005. 2005. General Requirements for the Competence of Testing and Calibration Laboratories, ISO. Geneva.

Keith, L. H. 1991. *Environmental Sampling and Analysis: A Practical Guide.* CRC: Boca Raton.

Koëter, H. B. W. M. 2003. Mutual acceptance of data: harmonized test methods and quality assurance of data—the process explained. *Toxicology Letters* 140–141: 11–20.

Kuttatharmmakul, S., Massart, D. L., and Smeyers-Verbeke, J. 2000. Comparison of alternative measurement methods: determination of the minimum number of measurements required for the evaluation of bias by means of interval hypothesis testing. *Chemometrics and Intelligent Laboratory Systems* 52: 61–73.

Lambert, D. H., Maher, W., and Hogg, I. 1992. Changes in phosphorus fractions during storage of lake water. *Water Research* 26: 645–648.

Meijer, S. N., Harner, T., Helm, P. A., Halsall, C. J., Johnston, A. E., and Jones, K. C. 2001. Polychlorinated naphthalenes in UK soils: time, trends, markers of source, and equilibrium status. *Environmental Science & Technology* 35: 4205–4213.

Merry, J. 1995. Quality management in wastewater treatment plants and reference materials for monitoring of nutrients in sea water environment. The approach, preparation, certification, and their use in environmental laboratories. *Fresenius Journal of Analytical Chemistry* 352: 148–151.

Morse, J. W., Hunt, M., Zulling, J., Mucci, A., and Mendez, T. 1982. A comparison of techniques for preserving dissolved nutrients in open ocean seawater samples. *Ocean Sciences and Engineering* 7: 75–106.

OECD. 1986. *Principles of Good Laboratory Practice.* In: OECD, Guidelines for Testing of Chemical, Annex. 2, OECD, Paris.

Ortiz-Estarelles, O., Martín-Biosca, Y., Medína-Hernández, M. J., Sagrado, S., and Bonet-Domingo, E. 2001. On the internal multivariate quality control of analytical laboratories. A case study: the quality of drinking water. *Chemometrics and Intelligent Laboratory Systems* 56: 93–103.

Päpke, O., Fürst, P., and Herrmann, T. 2004. Determination of polybrominated diphenylethers (PBDEs) in biological tissues with special emphasis on QC/QA measures. *Talanta* 63: 1203–1211.

Popek, E. P. 2003. *Sampling and Analysis of Environmental Chemical Pollutants*. Academic Press: Boston.

Ryden, J. C., Syers, J. K., and Harris, R. F. 1972. Sorption of inorganic phosphorus by laboratory ware. *Analyst* 97: 903–908.

Scholefield, D., Le Goff, T., Braven, J., Ebdon, L., Long, T., and Butler, M. 2005. Converted diurnal pattern in riverine nutrient concentrations and physical conditions. *Science of the Total Environment* 344: 201–210.

Silva, A. R. M. and Nogueira, J. M. F. 2008. New approach on trace analysis of triclosan in personal care products, biological and environmental matrices. *Talanta* 74: 1498–1504.

Smith, P. L. 2004. Audit and assessment of sampling systems. *Chemometrics and Intelligent Laboratory Systems* 74: 225–230.

Van der Veen, A .M. H., Linsinger, T. P. J., Lamberty, A., and Pauwels, J. 2001. Uncertainty calculations in the certification of reference materials. *Accreditation and Quality Assurance* 6: 257–263.

Vesely, J. 1990. Stability of the pH and the contents of ammonium and nitrate in precipitation samples. *Atmospheric Environment* 24A: 3085–3089.

Wagner, G. 1995. Basic approaches and methods for quality assurance and quality control in sample collection and storage for environmental monitoring. *The Science of the Total Environment* 176: 63–71.

Worsfold, P. J., Gimbert, L. J., Mankasingh, U., Omaka, O. N., Hanrahan, G., Gardolinski, P. C. F. C., Haygarth, P. M., Turner, B. L., Keith-Roach, M. J., and McKelvie, I. D. 2005. Sampling, sample treatment and quality assurance issues for the determination of phosphorus species in natural waters and soils. *Talanta* 66: 273–293.

Zhang, C. 2007. *Fundamentals of Environmental Sampling and Analysis*. Hoboken: Wiley-Interscience.

Zhang, J.-Z., Berberian, G., and Wanninkhof, R. 2002. Long-term storage of natural water samples for dissolved oxygen determination. *Water Research* 36: 4165–4168.

4 Experimental Design and Optimization Techniques

4.1 SYSTEM THEORY

Understanding the basic statistical models and the underlying theory of *experimental design* is crucial to employing such methods in research-based endeavors. Well-planned and effectively executed designs can maximize the amount of useful information obtained from a given study, especially with regard to dynamical environmental systems. The underlying theory of experimental design is depicted in the simplified system model presented in Figure 4.1a. A system under study is influenced by an *input* (quantity or quality that may have an influence upon the system) resulting in a defined, measurable *output*. Such a system can be viewed algebraically (Deming and Morgan 1993):

$$y = x + 2 \tag{4.1}$$

where x is the *independent variable* and y is the *dependent variable*. The equation transforms a given value of x into an output value, y. In the case of Equation 4.1, $x = 0$ for obtaining $y = 2$. These variables can be further defined as depicted in Figure 4.1b. Investigators can choose controlled *factors* (e.g., temperature, time) as input variables whose effects can be easily varied and evaluated by an output response(s). The investigator is also faced with the possibility of uncontrolled factors that may be *discrete* (e.g., the use of different instruments) or *continuous* (e.g., humidity) in character. For the latter, efforts to minimize these factors statistically through *randomization*, for example, are often employed (Section 4.3).

A *factor* of an experiment is a controlled independent variable—a variable whose *levels* are set by the investigator based on experience, previous studies, and intuition.

4.2 REVIEW OF LINEAR REGRESSION MODELS AND MATRIX NOTATION

Chapter 2 discusses the importance of expressing the results of an experiment in quantitative terms. This is facilitated by the use of *linear regression* models that

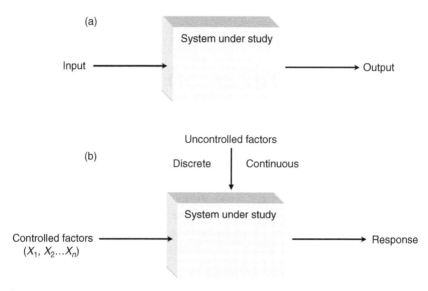

FIGURE 4.1 System model of experimental design depicting: (a) a simplified system influenced by input variables resulting in a defined output and, (b) further definition of the system including independent and dependent variables and measured response.

show the relationship between the dependent variable (y) and the independent variables x_1, x_2, $x_3 \ldots x_k$ (often termed k *independent* or *regressor variables*), discussed in Section 4.1. Consider the development of a spectrophotometric method for the determination of analyte X in natural waters. We wish to construct a model relating the obtained absorbance measurements to a pH and reagent concentration. The generalized model describing this relationship is

$$y = \beta_0 + \beta_1 x_1 + \beta_2 x_2 + \varepsilon \tag{4.2}$$

where y is the absorbance, x_1 the pH and x_2 is the reagent concentration. The term β_0 defines the intercept of the plane of the two-dimensional x_1, x_2 space. β_1 and β_2 are *partial regressor coefficients*, with β_1 measuring the change in y per unit change in x_1 (with x_2 held constant), and β_2 measuring the change in y per unit change in x_2 (with x_1 held constant). The term ε represents random experimental error. When considering k regressor variables the model can be shown as (see Section 6.8 for greater discussion on multiple linear regression)

$$y = \beta_0 + \beta_1 x_1 + \beta_2 x_2 + \cdots + \beta_k x_k + \varepsilon \tag{4.3}$$

What if we add an interaction term to this first-order model presented? It now becomes slightly more complex and can be written as

$$y = \beta_0 + \beta_1 x_1 + \beta_2 x_2 + \beta_{12} x_1 x_2 + \varepsilon \tag{4.4}$$

Here, the y is the response for given levels of the main effects x_1 and x_2 with the $x_1 x_2$ term accounting for the possible interaction effect between x_1 and x_2. Equations 4.3

and 4.4 can be used to express the results of two-level factorial designs quantitatively (Section 4.5.1). *Second-order models* include main effects, interactions, and quadratic effects and are typically used in *response surface* modeling. Considering two variables, the equation can be defined as

$$y = \beta_0 + \beta_1 x_1 + \beta_2 x_2 + \beta_{11} x_1^2 + \beta_{22} x_2^2 + \beta_{12} x_1 x_2 + \varepsilon \tag{4.5}$$

Estimation of the regression coefficients in such models is typically done by the use of the method of least-squares, and involves the generation of advanced least-squares normal equations. Generally, the *matrix notation* provides a simpler, more compact way of performing computations solving regression problems. Basic concepts and definitions of matrix theory are reviewed in Appendix III. For regression, consider the following equation

$$\mathbf{Y} = \mathbf{X}\boldsymbol{\beta} + \boldsymbol{\varepsilon} \tag{4.6}$$

where $\mathbf{Y} = n \times 1$ vector of observations, $\mathbf{X} = n \times p$ matrix of the levels of independent variables, $\boldsymbol{\beta} = p \times 1$ vector of the regression coefficients, and $\boldsymbol{\varepsilon} = n \times 1$ vector of random errors. If broken down into individual components we have

$$\mathbf{Y} = \begin{bmatrix} y_1 \\ y_2 \\ \vdots \\ y_n \end{bmatrix} \quad \mathbf{X} = \begin{bmatrix} 1 & x_{11} & x_{12} & \cdots & x_{1k} \\ 1 & x_{21} & x_{22} & \cdots & x_{2k} \\ \vdots & \vdots & \vdots & & \vdots \\ 1 & x_{n1} & x_{n2} & \cdots & x_{nk} \end{bmatrix} \quad \boldsymbol{\beta} = \begin{bmatrix} \beta_0 \\ \beta_1 \\ \vdots \\ \beta_k \end{bmatrix} \quad \boldsymbol{\varepsilon} = \begin{bmatrix} \varepsilon_0 \\ \varepsilon_1 \\ \vdots \\ \varepsilon_n \end{bmatrix}$$

A simulated set of data relating to the design of the spectrophotometric method referred to here is presented in Table 4.1. From this set of eight experiments we can generate

$$\mathbf{X} = \begin{bmatrix} 1 & 4.6 & 10.5 \\ 1 & 5.0 & 10 \\ 1 & 5.3 & 11 \\ 1 & 4.2 & 10.5 \\ 1 & 3.5 & 9.5 \\ 1 & 5.0 & 10 \\ 1 & 4.5 & 11.5 \\ 1 & 4.2 & 12 \end{bmatrix} \quad \mathbf{Y} = \begin{bmatrix} 0.1044 \\ 0.1265 \\ 0.0989 \\ 0.1332 \\ 0.0955 \\ 0.1451 \\ 0.1099 \\ 0.0977 \end{bmatrix}$$

From this a least-squares estimate of $\boldsymbol{\beta}$ can be produced using the equation

$$\hat{\boldsymbol{\beta}} = (\mathbf{X}'\mathbf{X})^{-1} \mathbf{X}'\mathbf{Y} \tag{4.7}$$

incorporating the matrix multiplication rules highlighted in Appendix III. The least-squares fit with regression coefficients can then be reported.

TABLE 4.1

**Simulated Experimental Data for the Development of a
Spectrophotometric Method for the Determination of
Environmental Species in Natural Waters**

Experiment	pH	Reagent Concentration (g/L)	Absorbance (a.u.)
1	4.6	10.5	0.1044
2	5.0	10	0.1265
3	5.3	11	0.0989
4	4.2	10.5	0.1332
5	3.5	9.5	0.0955
6	5.0	10	0.1451
7	4.5	11.5	0.1099
8	4.2	12	0.0977

4.3   EXPERIMENTAL DESIGN CONSIDERATIONS

The main applications of experimental design typically explored include *factor screening, response surface examination, system optimization*, and *system robustness*. To implement such procedures properly the following steps should be considered:

1. Determination of the overall objectives of the experiment.
2. Definition of the outcome (response) of the experiment.
3. Definition of the factors (and their levels) that will influence the response of an experiment.
4. Selection of a design that is compatible with the overall objectives, number of factors considered, and required precision of measurements.

This is opposite to the classical univariate approach. Such methods are time-consuming (especially when multiple factors are being considered) to the extent that the response is investigated for each factor while all others are held at a constant level. This approach is relatively simple, and suitable for factors that are independent. However, univariate methods do not take into account the interactive effects among factors. If the effects are additive in nature, experimental designs are the optimum choice and require fewer measurements.

4.3.1   EXPERIMENTAL UNCERTAINTY AND REPLICATION

Section 2.1 shows that uncertainty in results can be minimized but never completely eliminated. This also applies to experimental design concepts. Investigators can use the process of replication (repetition for each factor combination) to allow a better estimate of experimental error, which helps determine whether the observed differences in the dataset are truly statistically different. Replication also allows an estimation of the true mean response for one or more factor levels, thus helping to define precisely the effect of a factor on the response.

4.3.2 SAMPLE SIZE AND POWER

The importance of sample size in the distribution of repeated measurements is stressed in Chapter 2. Selection of appropriate sample size is also important when designing experiments intended for testing statistical hypotheses. Investigators should consider both the desired level of significance (usually denoted as α) and the desired *power* of the test. Power relates to the probability of correctly rejecting H_0 when H_0 is false. Ostensibly, one can control both by the selection of the number of n replicates. In the case of the power, fixed α increases as n increases.

4.3.3 BLOCKING AND RANDOMIZATION

Blocking is one of the fundamental principles of good experimental design and is employed when an investigator is aware of the presence of extraneous sources of variation that could influence the response. The blocking process itself reduces the variability from the most important sources and hence increases the precision of experimental measurements. Essentially, experimental units are grouped into homogeneous clusters in an attempt to improve the comparison of treatments by randomly allocating them within individual clusters or *blocks*. This is particularly helpful, for example, when one has to perform experimental runs on two separate days. Here, a set of two blocks would be appropriate with any day-to-day variation likely removed. *Randomization* could then be used to reduce the variability from the remaining extraneous sources. Its purpose is to ensure that the layout of the experiment does not consistently favor one or the other treatment in such a way that results would be misleading.

4.3.4 ORTHOGONALITY

Orthogonality is an important property that can help make even complex designs compact in nature. Consider the preceding discussion on matrices. If every pair of the different rows housed in $\mathbf{X}$ in Equation 4.6 results in a product of zero, the rows are considered orthogonal. An experimental design is orthogonal if the effects of any factor balance out (sum = 0) across the effects of the other factors present. The greater importance of this is presented in the discussion of 2^k designs (Section 4.5.1).

4.3.5 CONFOUNDING

Confounding is the relationship between 2 or more variables, which prevents their effects from being evaluated separately, and is highly dependent on the specific experimental design. Factors or interactions are confounded when the design array is configured in such a way that the effect of one factor is combined with the other.

4.3.6 CENTER POINTS

Designs with factors set at two levels assume that the effect of the factors on the dependent variable of interest is linear. We learn from our discussion in subsequent sections that it is impossible to test whether there are nonlinear (e.g., quadratic) components in the relationship between a factor and the dependent variable, if the factor is only evaluated at a low, and a high, value. If the investigator suspects that the relationship between the factors in the design and the dependent variable is curve-linear, one or more runs are performed with all (continuous) factors set at their midpoint (0), or center point.

4.4 SINGLE FACTOR CATEGORICAL DESIGNS

4.4.1 RANDOMIZED BLOCK DESIGNS

It clearly states in Section 4.3.3 that blocking can be used to reduce or eliminate the experimental error contributed by nuisance factors. Visualize a randomized block design in Table 4.2 as a collection of completely randomized experiments (nine in total), each run within one of the blocks (three in total) of the total experiment. Within blocks, it is then possible to assess the effect of different levels (high and low) of the factor without worrying about variations (e.g., day-to-day experiments) due to changes of the block factors, which are accounted for in the analysis.

4.4.2 LATIN SQUARE DESIGN

A *Latin square* is a block design with the arrangement of v Latin letters into a $v \times v$ array (a table with v rows and v columns). Latin square designs are often used in experiments in which subjects are allocated treatments over a given time period, and time is thought to have a major effect on the experimental response. Suppose that the treatments are labeled A, B, and C. In this particular situation, the design would be

---

TABLE 4.2
Collection of Completely Randomized Block Experiments

Experiment	Random Block	X_1	X_2	X_3
1	1	-1	-1	-1
2	1	1	1	-1
3	1	1	-1	1
4	2	-1	1	1
5	2	1	-1	-1
6	2	-1	1	-1
7	3	-1	-1	-1
8	3	1	1	1
9	3	-1	1	-1

Day 1 A B C

Day 2 C A B

Day 3 B C A

This type of design allows the separation of an additional factor from an equal number of blocks and treatments. If there are more than three blocks and treatments, a number of Latin square designs are possible. Consider the following four-treatment example.

EXAMPLE PROBLEM 4.1

1. Consider an analysis of the effect of different DO levels (treatment) in four experimental aquatic chambers on the % survival of a fish species. The four treatments were: 6.7%, 8.2%, 3.6%, and 5.9%. Develop an appropriate Latin square design.

Answer:

Experiment	Chamber 1 (%)	Chamber 2 (%)	Chamber 3 (%)	Chamber 4 (%)
1	6.7	8.2	3.6	5.9
2	8.2	3.6	5.9	6.7
3	3.6	5.9	6.7	8.2
4	5.9	6.7	8.2	3.6

From Example Problem 4.1, it should be obvious that a total of four basic experiments must be conducted. Knowing this, one can create an example design as shown in the following sections. Note that all treatments appear in a row or column only once. Latin square designs assume that there are no interactions among treatments, and are useful where there are no randomization restrictions (Montgomery 2005). It should be kept in mind that without replication, there is no pure way of estimating error. ANOVA models can be generated in terms of treatment, row, and column totals. If the Latin square design is replicated, the model is similar, but with an additional term for error produced.

4.4.3 GRECO-LATIN SQUARE DESIGN

The Greco-Latin square design involves two Latin squares that are superimposed on each other. It contains two treatment factors instead of one, and four factors overall instead of three. An example design would look like this:

	A_1	A_2	A_3	A_4
B_1	C_1D_3	C_2D_4	C_3D_1	C_4D_2
B_2	C_4D_2	C_1D_1	C_2D_3	C_3D_4
B_3	C_3D_1	C_4D_2	C_1D_3	C_2D_4
B_4	C_2D_4	C_1D_3	C_3D_2	C_4D_1

The analysis for the Greco-Latin square design is similar to that of a Latin square design. However, a noticeable difference is that two treatment sum of squares have to be computed (factors C and D) by listing two sets of means outside the design table. Note that Greco-Latin squares are most effective if replicated, and are subject to the same randomization rules as for the Latin square design.

4.5 SCREENING DESIGNS

A *screening* experiment is a systematic approach to identifying the key input parameters of a process or product that affects the output performance, especially when a large number of factors are present, as is the case in many environmental analyses. It uses a fraction of the total possible combinations of the potential inputs to determine which ones are the most important. Follow-up experiments, such as those discussed in Section 4.6, can then be performed to study their effects and possible interactions more thoroughly.

4.5.1 FULL FACTORIAL DESIGNS (TWO LEVELS PER FACTOR)

The most general two level design is full factorial and can be described as the 2^k type, where the base 2 stands for the number of factor levels and k, the number of factors each with an assigned low and high value. In a full factorial design, the levels of the factors are chosen in such a way that they span the complete factor space. With two factors this defines a square in the factor space and with three factors, a cube. The method can be graphically illustrated by a simplified example:

EXAMPLE PROBLEM 4.2

1. Consider the investigation of the effects of reaction temperature and pH in determining the spectrophotometric response (absorbance) of a standard analyte solution.

Answer: Figure 4.2 shows a graphical definition of the experimental domain with the reaction temperature varying from 50°C (low level) to 70°C (high level), and the reaction pH varying from 2 (low level) to 5 (high level). The best experimental points in the domain are located in the corners A, B, C, and D as follows:

A (50°C, pH 2); B (70°C, pH 2); C (50°C, pH 5); D (70°C, pH 5)

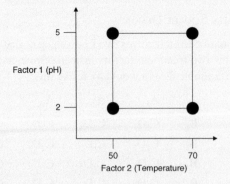

FIGURE 4.2 Graphical definition of the experimental domain with the reaction temperature varying from 50°C (low level) to 70°C (high level) and the reaction pH varying from 2 (low level) to 5 (high level).

TABLE 4.3

Experimental Matrix for the Spectrophotometric Response of a Standard Analyte Solution

Experiment	Temperature (°C)	pH	Absorbance (a.u.)[a]
A	70	5	0.1004
B	50	5	0.0856
C	70	2	0.1098
D	50	2	0.0944
Factor Levels			
(−)	50	2	
(+)	70	5	

[a] a.u. = arbitrary units.

The four trials of the experimental matrix used here are shown in Table 4.3, with the result of each experiment indicated in the response column, and the factor levels shown in the rows below the experimental matrix. Note that −1 is used for the low level of each factor and +1 for the high level. If we introduce another variable (e.g., reagent concentration) into the experiment, it then becomes possible to represent the factors as faces on one or more cubes with the responses at the points. The distribution of experimental points within this type of experimental domain (2^3 design) is shown schematically in Figure 4.3.

Considering the experimental design matrix in Table 4.3, we can calculate the main effects as well as interactive effects for the investigation. The main effects of factor temperature and pH can be calculated as follows:

$$\text{Effect}_{\text{Temp}} = \frac{0.1004 + 0.1098}{2} - \frac{0.0856 + 0.0944}{2} = 0.0151$$

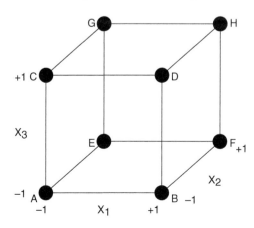

FIGURE 4.3 Schematic distribution of experimental points within the experimental domain of a 2^3 design.

$$\text{Effect}_{pH} = \frac{0.1004 + 0.0856}{2} - \frac{0.1098 + 0.0944}{2} = -0.009$$

As calculated, the effect of temperature is positive, suggesting to the investigator that increasing the temperature from 50°C to 70°C would result in a higher absorbance reading. In colorimetric analyses, the amount (intensity) of color developed as a result of an optimized reaction is measured by a spectrophotometer. As the color increases, the absorbance also increases at the specified wavelength. The effect of pH is negative, suggesting that increasing the pH would decrease the resultant absorbance reading.

Main effect is the difference between the average responses of all observations in a given factor's high and low levels.

Interactive effect is a second-order effect based on the joint effect of two single factors.

How do the manual calculations compare with the use of statistical software in performing a 2^2 factorial design? What extra information can be gained? Here, the basic calculations for the linear model can be shown in an ANOVA analysis, and the significance of each factor on the response (absorbance) can be assessed by examination of Prob $> F$ from the Effect Test results.

The basic calculations for the linear model are shown in the ANOVA analysis (Table 4.4). As discussed in Chapter 2, ANOVA for a linear regression partitions the total variation of a sample into various components, which are used to compute an F-ratio that evaluates the effectiveness of the model. Prob $> F$ is the significance probability for the F-ratio, which states that if the null hypothesis were true, a larger F-statistic would only occur due to random error. Significance probabilities of 0.05 or less are often considered evidence that there is at least one significant regression factor in the model. An examination of Prob $> F$ from the Effect Test results (Table 4.5) reveals the parameter estimates on the response. Although both temperature and pH has statistically significant effects on absorbance, temperature has the greatest, at a Prob $> F = 0.0126$. This agrees with our results using the preceding manual calculations.

Examination of the interaction effect between temperature and pH is a bit problematic, as good experimental designs normally require more than two factors. However, we can create an additional table (Table 4.6) representing the new design

---

TABLE 4.4

ANOVA Table for the Linear Model

Source	DOF	Sum of Squares	Mean Square	F-Ratio
Model	2	0.00031082	0.000155	1726.778
Error	1	0.00000009	0.00000008	Prob $> F$
C. Total	3	0.00031091		0.0170

TABLE 4.5

Effect Test for the Factors Affecting Absorbance Readings

Source	Number of Parameters	DOF	Sum of Squares	F-Ratio	Prob > F
Temperature	1	1	0.00022801	2533.444	0.0126
pH	1	1	0.00008281	920.1111	0.0210

TABLE 4.6

Experimental Matrix for the Spectrophotometric Response of a Standard Analyte Solution Including the Interaction Effect

Experiment	Temperature	pH	Temperature/ pH Interaction	Absorbance (a.u.)[a]
A	+1	+1	+1	0.1004
B	−1	+1	−1	0.0856
C	+1	−1	−1	0.1098
D	−1	−1	+1	0.0944

[a] a.u. = arbitrary units.

matrix with the interaction included. The interaction effect can then be calculated as follows:

$$\text{Effect}_{\text{interaction}} = \frac{0.1004 + 0.0944}{2} - \frac{0.0856 + 0.1098}{2} = -0.003$$

which is smaller than the main effect for both temperature and pH. This can be visually demonstrated with *interaction plots* using the data from the given example. In an interaction plot, evidence of interaction shows as nonparallel lines. As can be seen in Figure 4.4, these lines are parallel, suggesting that the interaction is very small

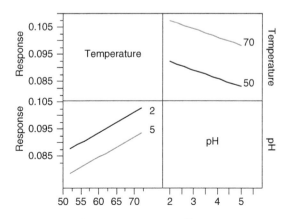

FIGURE 4.4   Interaction plot of temperature versus pH.

(as shown in the manual calculations above) and separating the effects is thus appropriate.

EXAMPLE PROBLEM 4.3

1. Consider Figure 4.3. Produce an experimental design matrix for the full factorial (2^3) design depicted using $y_1, y_2, y_3, \ldots, y_k$ as the response.

Answer:

TABLE 4.7

Experimental Design Matrix for the Full Factorial (2^3) Design

Experiment	x_1	x_2	x_3	Response
1	-1	-1	-1	y_1
2	$+1$	-1	-1	y_2
3	$+1$	$+1$	-1	y_3
4	-1	$+1$	-1	y_4
5	-1	-1	$+1$	y_5
6	$+1$	-1	$+1$	y_6
7	-1	$+1$	$+1$	y_7
8	$+1$	$+1$	$+1$	y_8

4.5.2 FRACTIONAL FACTORIAL DESIGNS (TWO LEVELS PER FACTOR)

In *fractional factorial designs*, the number of experiments is reduced by a number p according to a 2^{k-p} design. In the most commonly employed fractional design, that is, the half-fraction design ($p = 1$), exactly one half of the experiments of a full design are performed. For a closer look at a two-level fractional factorial design consider the following example:

EXAMPLE PROBLEM 4.4

1. Describe a 2^{3-1} design.

Answer: For the 2^{3-1} design above, we can deduce that it is one with three factors and four treatment combinations. Ostensibly, this design is numerically equivalent to a 2^2 design. The 2^{3-1} can also be termed a *half replicate* of a 2^3 design as we can write $2^3 \cdot 2^{-1} = 2^3 \cdot \frac{1}{2}$.

The fractional factorial design is based on an algebraic method of calculating the contributions of the numerous factors to the total variance, with less than a full factorial number of experiments. Such designs are useful when the number of potential factors is relatively large ($n \geq 7$) because they reduce the total number of runs required for the overall experiment.

So far, discussions have centered on two-level screening designs, with limited coverage of three-level designs. Now consider *mixed-level designs*, those incorporating, for example, combinations of two and three levels. Take an experiment for the analysis of selective serotonin reuptake inhibitor drugs in environmental samples using solid-phase microextraction gas chromatography-mass spectrometry (Lamas et al. 2004). Here, a mixed-level $3 \times 2^{4-2}$ fractional factorial design was incorporated to study the influence of five experimental factors affecting extraction efficiency. The experimental matrix for this mixed-level design is shown in Table 4.8. All factors were studied at two levels with the exception of the amount (%) of NaCl added, which was studied at three levels.

In this study, the statistical significance for the effects was determined by the use of standardized first-order *Pareto charts* (Figure 4.5). Pareto charts are specialized versions of histograms (Chapter 2) that highlight the most significance factors in increasing order. As shown, NaCl is only significant for sertraline when the low and high values are compared. Two other factors, volume of derivatization and thermostatization time, for example, are not significant.

4.5.3 PLACKETT–BURMAN AND TAGUCHI DESIGNS

The earlier discussions focused on how using fractional factorial designs can be instrumental in obtaining a substantial amount of information in a reduced number

---

TABLE 4.8

Experimental Matrix for the Mixed-Level Fractional Factorial Design

Experiment	NaCl (%)	Extraction Temperature (°C)	Derivatization Reagent (µL)	Thermostatization Time (Min)	Fiber Coating[a]
1	0	50	40	5	PDMS-DVB
2	0	50	80	15	PDMS-DVB
3	0	100	40	5	CW-DVB
4	0	100	80	15	CW-DVB
5	15	50	40	5	CW-DVB
6	15	50	80	15	CW-DVB
7	15	100	40	5	PDMS-DVB
8	15	100	80	15	PDMS-DVB
9	30	50	40	5	PDMS-DVB
10	30	50	80	15	PDMS-DVB
11	30	100	40	5	CW-DVB
12	30	100	80	15	CW-DVB

[a] PDMS-DVB, polydimethylsiloxane-divinylbenzene; CW-DVB, carbowax-divinylbenzene.

Source: Data from Lamas, J. P., et al., *Journal of Chromatography A*, 1046, 241–247, 2004. With permission from Elsevier.

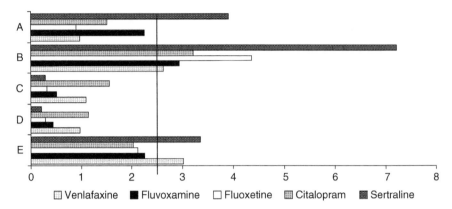

FIGURE 4.5   Pareto plot used to test statistical significance in the analysis of selective serotonin reuptake inhibitor drugs via solid-phase microextraction gas chromatography–mass spectrometry. (From Lamas, J. P., et al., *Journal of Chromatography A*, 1046, 241–247, 2004. With permission from Elsevier.)

of experiments. However, some interactions and factors are *aliased* in such designs, where the main effect and interaction cannot be individually analyzed. Such is the case in the 2^{3-1} design just presented, where factor C is aliased with the AB interaction (Table 4.9).

Plackett–Burman screening (resolution III) designs overcome this limitation, though at the expense of partial confounding. Here, full factorial designs can be fractionalized to yield saturated designs where the number of experiments is a multiple of four rather than a multiple of 2 (2^k). In saturated designs, the information obtained is used to estimate the factors, leaving no degrees of freedom to estimate the error term for the ANOVA. *Plackett–Burman designs* can be constructed from *Hadamard matrices* of the order N, where there is an $N \times N$ orthogonal matrix with entries 1 (+) or −1 (−) (Wu and Hamada 2000). There are no two-level fractional factorial designs with sample sizes between 16 and 32 runs. However, there are 20-, 24-, and 28-run Plackett–Burman designs. The generating vectors for selected Plackett–Burman designs have $n - 1$ rows. For Plackett–Burman designs, the main effects are orthogonal; and two-factor interactions are only partially confounded

TABLE 4.9

2^{3-1} Design Where Factor C Is Aliased with the AB Interaction

Experiment	A	B	AB + C
1	−1	−1	+1
2	+1	−1	−1
3	−1	+1	−1
4	+1	+1	+1

with main effects. This is in contrast to resolution III fractional factorial where two-factor interactions are indistinguishable from main effects.

For *resolution III*, the main effects are confounded with one or more two-way interactions, which must be assumed to be zero for the main effects to be significant.

Consider a two-level, 12-run, eight-factor Plackett–Burman used by Moreda-Piñeiro et al. (2002) in their study of mercury (Hg) determination of environmental and biological samples by cold vapor generation–electrothermal atomic absorption spectrometry. More detailed theory behind the cold vapor generation is presented in López-García et al. (1991). The different factors affecting the Hg cold vapor generation efficiency and their levels selected in this study are provided in Table 4.10. Can you picture an example of what the design matrix would look like? The 12-run, fully generated, experimental matrix used in this study is shown in Table 4.11. Does this agree with your example? For this study, the main effect and two-factor Pareto charts were used to assess the significance of atomization and trapping temperatures, trapping time, and hydrochloric acid. A *central composite design* (Section 4.6.1) was used to generate optimum values from the significant factors.

Taguchi designs employ the use of *orthogonal arrays* to aid the study of the entire factor space and determine the optimal design within a limited number of experiments. Orthogonal arrays (designated L_n arrays, $N \times m$ matrix) are systematically constructed tables that are essentially two-level, three-level, and mixed-level fractional factorial designs. The columns in the L_n arrays are highly confounded. They are generally used as screening designs, with only selected interactions being investigated. Unique to this approach is the use of signal and noise factors, inner and outer arrays, and signal-to-noise

TABLE 4.10
Experimental Factors for the Plackett–Burman Design for Hg Cold Vapor Generation

Factor	Low (−)	High (+)
[HCl] (M)	0.5	1.0
[NaBH$_4$] (%)	1.0	2.0
Air flow rate (mL/s)	30	60
Trapping temperature (°C)	200	800
Atomization temperature (°C)	2000	2500
Trapping time (s)	20	40
Particle size (μM)	2	50
Acid solution volume (mL)	2.5	5.0

Source: Data from Moreda-Piñeiro, J., et al., *Analytica Chimica Acta*, 460, 111–122, 2002. With permission from Elsevier.

TABLE 4.11

Plackett–Burman Design Matrix for the 12 Run Hg Cold Vapor Generation Experiment

Runs	[HCl] (M)	[NaBH$_4$] (%)	Air flow Rate (mL/s)	Trapping Temperature (°C)	Atomization Temperature (°C)	Trapping Time (s)	Particle Size (μM)	Acid Solution Volume (mL)
1	+	−	+	−	−	−	+	+
2	+	+	−	+	−	−	−	+
3	−	+	+	−	+	−	−	−
4	+	−	+	+	−	+	−	−
5	+	+	−	+	+	−	+	−
6	+	+	+	−	+	+	−	+
7	−	+	+	+	−	+	+	−
8	−	−	+	+	+	−	+	+
9	−	−	−	+	+	+	−	+
10	+	−	−	−	+	+	+	−
11	−	+	−	−	−	+	+	+
12	−	−	−	−	−	−	−	−

Source: Data from Moreda-Piñeiro, J., et al., *Analytica Chimica Acta*, 460, 111–122, 2002. With permission from Elsevier.

(S/N) ratios. Noise factors are described as those that cannot be controlled in normal operations, but can be controlled for the purpose of testing.

Taguchi designs use an inner design constructed over the control factors to aid the search for the optimum settings. An outer design over the noise factors looks at how the response behaves for a wide range of possible noise conditions. The experiment is performed on all combinations of the inner and outer design runs, and a performance statistic is calculated across the outer runs for each inner run. This becomes the response for a fit across the inner design runs. If the goal is to achieve a target response, for example, then maximizing the S/N ratio will give the factor settings that achieve the target response with the least variation.

Taguchi designs can be used to assign two-level factors to one of the series of orthogonal arrays such as L$_8$ (2^7), L$_{16}$ (2^{15}), and L$_{32}$ (2^{31}). Assigning three-level factors to orthogonal arrays is also appropriate, and commonly done [e.g., L$_9$ (3^4)]. In such arrays, Taguchi designated the levels as 1 (low) and 2 (high). Consider this example problem.

Using Taguchi's orthogonal arrays allows a simultaneous estimation of all the effects with a relatively simple and manageable experimental design. ANOVA can be used to analyze the effects of the factors on the response. Both the response and S/N ratios can be used to find the optimum conditions.

What about situations where the investigator cannot assume that all interactions are zero? Linear graphs (two alternatives for L$_8$ are depicted in Figure 4.6) are used to help allocate the appropriate columns for specific interactions to be estimated.

EXAMPLE PROBLEM 4.5

1. Construct a generalized Taguchi L_8 orthogonal array table.

Answer: Begin by noting that the arrays are organized so that the factors are placed across the top of the table (orthogonal array) as actual column headings. Ostensibly, the rows correspond to treatment combinations. *Important: the number of columns in a two-level design array is one less than the number of rows, and corresponds to the number of degrees of freedom.* For L_8 (2^7), we have eight rows and seven columns arranged as follows:

TABLE 4.12

A Generalized Taguchi L_8 Orthogonal Design Matrix

Experiment Number	Factor 1	Factor 2	Factor 3	Factor 4	Factor 5	Factor 6	Factor 7
1	1	1	1	1	1	1	1
2	1	1	1	2	2	2	2
3	1	2	2	1	1	2	2
4	1	2	2	2	2	1	1
5	2	1	2	1	2	1	2
6	2	1	2	2	1	2	1
7	2	2	1	1	2	2	1
8	2	2	1	2	1	1	2

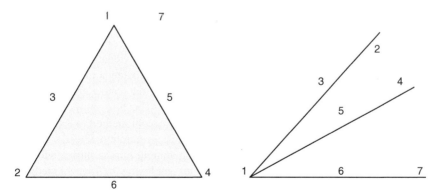

FIGURE 4.6 Linear graphs (two alternatives for L_8) used to help allocate the appropriate columns for specific interactions to be estimated. Numbers refer to actual columns of the orthogonal array and the number on the line connecting two vertices represents the interaction between the vertices.

TABLE 4.13

L₉ Orthogonal Arrays Applied in Optimizing a Reverse Osmosis Wastewater Treatment Process

Experiment Number	Temperature (°C)	Pressure (Bar)	Concentration (ppm)
1	1	1	1
2	1	2	2
3	1	3	3
4	2	1	2
5	2	2	3
6	2	3	1
7	3	1	3
8	3	2	1
9	3	3	2

Source: Data from Madaeni, S. S. and Koocheki, S., *Chemical Engineering Journal*, 119, 37–44, 2006. With permission from Elsevier.

Here, the numbers refer to actual columns of the orthogonal array, and the number on the line connecting the two vertices represents the interaction between the vertices. For example, column 3 contains the interaction between columns 1 and 2.

Recently, an L₉ Taguchi method was applied to optimize a reverse osmosis wastewater treatment process in an industrial setting (Madaeni and Koocheki 2006). The main objective of this study was to find optimum conditions to achieve high flux and high rejection in terms of reverse osmosis membranes for treatment of water containing unwanted levels of nitrate, nitrite, sulfite, and phosphate. Factors such as transmembrane pressure, temperature, and concentration affect the flux and rejection, and were therefore chosen and investigated at three levels for this study.

The L₉ orthogonal array used is provided in Table 4.13. Here, the levels for the three factors were: temperature ($1 = 25 \pm 2°C$, $2 = 30 \pm 2°C$, $3 = 35 \pm 2°C$); pressure ($1 = 5.75 \pm 0.2$ bar, $2 = 6.25 \pm 0.2$ bar, $3 = 6.75 \pm 0.2$ bar); concentration ($1 = 50 \pm 5$ ppm, $2 = 80 \pm 5$ ppm, $3 = 110 \pm 5$ ppm). For the latter, individual (and combined ion) solutions of nitrate, nitrite, sulfite, and phosphate were prepared at the levels listed for all flux measurements (l/m^2 h). ANOVA results indicate that temperature and pressure have the most significant contribution to water flux. The flux shows a substantial improvement from 58 l/m^2h to 69 l/m^2h using the Taguchi method to set the control factor conditions. This technique also shows that the concentration of feed solution has the highest significance in rejecting a solution containing nitrate, nitrite, sulfite, and phosphate. Rejection was enhanced to 99.9% after employing the Taguchi method.

4.6 THREE-LEVEL DESIGNS: RESPONSE SURFACE METHODOLOGY

Much of the discussion in previous sections centers on the dependence of a response on specific levels of independent factors. As we learn from Section 4.2, however, *response surface methodology* (RSM) is useful for modeling a curved quadratic surface to continuous factors. This approach is used when simple linear and inter-action models are not adequate, for example, when the experimentation is far from the region of optimum conditions. Here, the investigator can expect curvature to be more prevalent and will need a mathematical model that can represent the curvature. We also learn that second-order models include the main effect, interactions, and quadratic effects with the simplest two-factor model represented by Equation 4.5. Curvature can be modeled if the factor levels are investigated at least three levels. Three-level factorial designs are thus known as response surface designs. There are entire books on RSM that describe this approach in great detail. Therefore, we present only some of the most common RSM techniques employed in environmental analysis, the basic theory behind them, and modern applications utilizing such techniques.

4.6.1 CENTRAL COMPOSITE DESIGNS

One of the most popular response surface designs utilized in environmental analysis is the central composite design (CCD), with a two-factor example illustrated in Figure 4.7. Central composite designs contain embedded factorial or fractional facto-rial designs with center points that are augmented with a group of *axial* (star) runs that allow estimation of curvature. Consider a CCD that consists of a full factorial, two-level design with an embedded star design. How can the number of runs be determined? We can obtain this value, r, by (Otto 1999)

$$r = 2^{k-p} + 2k + n_0 \tag{4.8}$$

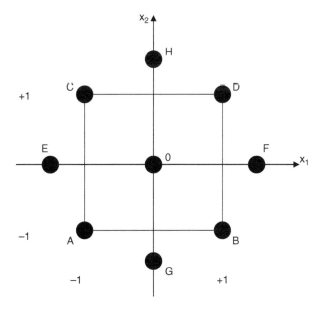

FIGURE 4.7 Schematic of a two-factor central composite design.

where k is the number of factors, p the number of reductions of the full design, and n_0 is the number of experiments in the center of the design. Central composite designs consist of cube points at the corners of a unit cube that is the product of the intervals $[-1, 1]$, star points along the axes at or outside the cube, and center points at the origin. Points A, B, C, and D are the points of the initial factorial design, with points E, F, G, and H as the star points at the central 0.

Three distinct types of CCD are used together with second-order response models, with each depending on where the star points are located. In circumscribed central composite circumscribed (CCC) designs the star points are at some distance, α, from the center, primarily based on the number of factors in the design. These star points in turn establish new extremes for the low and high settings for all chosen factors . Such designs require five levels and can be produced by augmenting an existing factorial or from the resolution V fractional factorial design. Like the CCC, an inscribed central composite inscribed (CCI) requires five levels of each factor. Here, the factorial points are brought into the interior of the design space (thus considered inscribed) and set at a distance from the center point that maintains the proportional distance of the factorial points to the axial points. Ostensibly, this is a limited version of the CCC, and also a convenient way to generate a rotatable CCD that enables the investigator to study the full ranges of the experiment factors while excluding the conditions atone or more of the extremes of the experimental design region. In a central composite face-centered design (CCF), three levels are needed for each factor, with the star points located at the center of each face of the factorial space. Like the CCC, this design can be produced by augmenting an existing factorial or the resolution V design.

EXAMPLE PROBLEM 4.6

1. How many experiments are there in a CCD with three factors?

Answer: Using equation 4.8 we obtain:

$$r = 2^3 + 2 * 3 + 1 = 15$$

Consider an application by Giokas and Vlessidis (2007) who incorporated a three-factor, five-level CCD (Figure 4.8) with two orthogonal blocks, for optimizing the determination of aqueous photolysis rates of two UV absorbing organic compounds in sunscreen products [octyl methoxy cinnamate (E2292) and 3-(4-methyl-benzyldene)-camphor (E6300)] in natural waters. The experimental design matrix (Table 4.14) consists of 17 experiments of which 14 are experimental points and three are star replications. Included in this table are photodegredation kinetic data and coefficients for ln (concentration after time t/initial concentration at t_0) versus time. Linear regression of the logarithmic concentration values determined as a function of time was used to generate pseudo-first-order reaction rate constants. Factor levels were chosen by the authors based on prior examination of the effect of each factor in various conditions ranging from clear to polluted water.

A post-hoc review of their model revealed that the maximum photodegredation of E6300 was accomplished at the following conditions: NO_3^- (19 mg L^{-1}), DOM

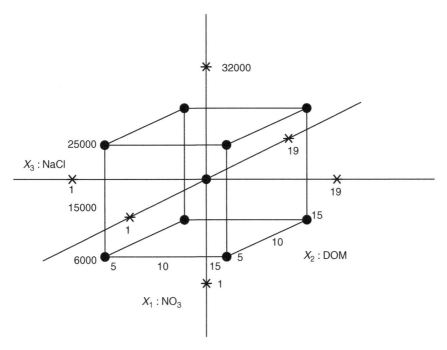

FIGURE 4.8 Three-factor, five-level CCD with two orthogonal blocks for optimizing the determination of aqueous photolysis rates of two UV absorbing organic compounds in sunscreen products [octyl methoxy cinnamate (E2292) and 3-(4-methylbenzyldene)-camphor (E6300)] in natural waters. (From Giokas, D. L. and Vlessidis, A. G., *Talanta*, 71, 288–295, 2007. With permission from Elsevier.)

(10 mg L^{-1}), and NaCl (1 mg L^{-1}). For E2292, the maximum photodegradation occurred at: NO$_3^-$ (19 mg L^{-1}), DOM (1 mg L^{-1}), and NaCl (15,000 mg L^{-1}). The coefficients of multiple determinations (r^2) representing the fit models were 0.90 and 0.97 for E2292 and E6300, respectively. What do these values represent in the model? Statistically, it means that 90 and 97% of the data about their mean were accounted for by the factor effects in the model. ANOVA analysis reveals that the linear and quadratic effects of NO$_3^-$ and the quadratic effect of DOM were statistically significant at the 95% level of confidence. All cross-product terms were significant for E6300. For E2292, only the linear effects of DOM and NO$_3^-$ were found to affect the reaction-rate constant. Results proved the effectiveness of a CCD in delineating factor significance in complex environmental matrices over a large concentration range.

4.6.2 Box–Behnken Design

The *Box–Behnken design* is considered an efficient option in RMS and an ideal alternative to central composite designs. It has three levels per factor, but avoids the corners of the space, and fills in the combinations of center and extreme levels (Figure 4.9). It combines a fractional factorial with incomplete block designs in such a way as to avoid the extreme vertices, and to present an approximately rotatable

TABLE 4.14
Central Composite Design for Optimizing the Determination of UV Absorbing Aqueous Photolysis Rates

Run	Block	NO$_3$ (mg L^{-1})	DOM (mg L^{-1})	NaCl (mg L^{-1})	K$_{photo}$ (E2292)	K$_{photo}$ (E6300)	r^2 (E2292)	r^2 (E6300)
1	1	5	5	6000	0.00732	0.00552	0.95	0.94
2	1	5	15	6000	0.00708	0.00782	0.97	0.95
3	1	5	5	25,000	0.0072	0.00449	0.97	0.96
4	1	5	15	25,000	0.00696	0.00552	0.99	0.97
5	1	15	5	6000	0.01164	0.00736	0.96	0.94
6	1	15	15	6000	0.01008	0.00840	0.92	0.96
7	1	15	5	25,000	0.01116	0.00966	0.95	0.97
8	1	15	15	25,000	0.00984	0.00782	0.94	0.99
9	1	10	10	15,000	0.0096	0.00817	0.99	0.98
10	1	10	10	15,000	0.00972	0.00794	0.95	0.99
11	2	10	1	15,000	0.01177	0.00713	0.98	0.99
12	2	10	19	15,000	0.00781	0.00690	0.97	0.98
13	2	10	10	1	0.00902	0.00874	0.99	0.94
14	2	10	10	32,000	0.00858	0.00805	0.97	0.97
15	2	1	10	15,000	0.00737	0.00518	0.96	0.95
16	2	19	10	15,000	0.01111	0.00978	0.95	0.98
17	2	10	10	15,000	0.00858	0.00851	0.92	0.96

Source: From Giokas, D. L. and Vlessidis, A. G., *Talanta*, 71, 288–295, 2007. With permission from Elsevier.

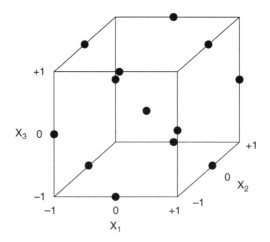

FIGURE 4.9 Schematic of a three factor, three-level Box–Behnken response surface design.

TABLE 4.15

Number of Factors and Experimental Runs for the Box–Behnken Design (Three Replicates in the Center of Each Design Is Assumed)

Factors Considered	Number of Experimental Runs
3	15
4	27
5	46
6	54
7	62

design with only three levels per factor. A design is rotatable if the variance of the predicted response at any point $\times$ depends only on the distance of $\times$ from the design's center point. It must be noted, however, that Box–Behnken designs should be confined to uses where the experimenter is not interested in predicting response at the extremes (corners of the cube). Compared to the CCD, the Box–Behnken design requires fewer experimental factor combinations with the number of experiments per number of representative factors listed in Table 4.15. An example design matrix for a three-factor Box–Behnken design is shown in Table 4.16. Listed in the table are the general coded levels and responses for all 15 experimental runs. Note the two replicates performed in the center of the design (0,0,0). Box–Behnken

TABLE 4.16

Box–Behnken Experimental Design Matrix for Three Factors

Run	x_1	x_2	x_3	Response
1	-1	-1	0	y_1
2	$+1$	-1	0	y_2
3	-1	$+1$	0	y_3
4	$+1$	$+1$	0	y_4
5	0	0	-1	y_5
6	0	0	$+1$	y_6
7	0	0	0	y_7
8	0	0	0	y_8
9	0	0	0	y_9
10	0	-1	0	y_{10}
11	0	$+1$	0	y_{11}
12	-1	0	-1	y_{12}
13	$+1$	0	-1	y_{13}
14	-1	0	$+1$	y_{14}
15	$+1$	0	$+1$	y_{15}

designs are appropriate for fitting a quadratic model based on the data generated. See the following environmental application example for an expanded look.

EXAMPLE PROBLEM 4.7

1. Can you name any disadvantages of using the Box–Behnken design in place of the CCD?

Answer: The Box–Behnken design contains regions of poor prediction quality despite being rotatable. The missing corner experiments may also provide a potential loss of data in the overall model.

Gfrerer and Lankmayr (2005) show the usefulness of a Box–Behnken experimental design in optimizing a microwave-assisted extraction (MAE) process for the determination of organochlorine pesticides in sediment samples. After extraction, sample analysis was carried out by gas chromatography combined with mass spectrometry (GC-MS). Due to the multitude of experimental factors present in such a study, it was necessary to perform an initial 2^6 factorial screening design, which shows that the composition of extraction solvent [acetone (vol. %)], extraction temperature (°C), and time [hold-up time (min)] are statistically significant. These factors were then assigned levels. The design matrix is presented in Table 4.17.

TABLE 4.17

Box–Behnken Design Matrix for Mae Determination of Persistent Organochlorine Pesticides

Run	Hold-Up Time (Min)	Temperature (°C)	Acetone (Vol.%)
1	45	150	40
2	45	115	20
3	45	115	60
4	25	115	40
5	25	115	40
6	5	115	60
7	25	150	20
8	5	115	20
9	25	150	60
10	5	150	40
11	5	80	40
12	25	80	60
13	25	115	40
14	45	80	20
15	45	80	40

Source: Modified from Gfrerer, M. and Lankmayr, E., *Analytica Chimica Acta*, 533, 203–211, 2005. With permission from Elsevier.

Maximum recovery (%) of each compound with respect to the values obtained by a previous Soxhlet extraction method was chosen as the target function for optimization.

ANOVA results (95% level of confidence) reveal extraction solvent as the most significant factor for most compounds tested. Extraction temperature was significant for all except α-endosulfan, pentochlorobenzene (PCBz), p,p'-DDD, and p,p'-DDT. The effect of extraction temperature was significant for PCBz and p,p'-DDT. What about the determination of optimum conditions? What techniques can investigators utilize to aid this process?

In order to determine the optimum ranges for the individual factors, they utilized the response surface function keeping one of the variables fixed at a central point. The authors used a main effect plot and response surface images (Figure 4.10) in their approach. The main effect plot shows how a maximum can be achieved for the factors within the specified ranges for the compound Aldrin. Optimized conditions for all compounds in this study proved to be the combination of 20 min extraction time at 120°C and a mixture of 1 + 1 (v/v) of n-hexane-acetone as the extraction solvent. Typical example response surfaces are shown (Figure 4.10) for central levels of (a) vol.% of acetone, (b) extraction temperature, and (c) extraction time for the MAE of Aldrin.

4.7   DOEHLERT MATRIX DESIGN

A very useful alternative design for second-order models is the Doehlert matrix design conceived by Doehlert in 1970 (Doehlert 1970). A comprehensive review by Ferreira et al. (2004) describes the experimental domain in detail and compares its advantages and disadvantages with the CCD and Box–Behnken designs. It also describes the spherical experimental domain and its uniformity in space filling. For two variables, for example, the design consists of a single central point and six points forming a regular hexagon. The levels are varied for each factor with one factor studied at five levels, for example, and the other at three. Which variable should be assigned at five levels? It should be obvious that the one suspected of having the most persuasive effect on the response must be chosen.

A major advantage of this design is that fewer experiments are needed since they have been shown to be extremely efficient as they move through the experimental domain (Bourguignon et al. 1993). Massart et al. (2003) discussed efficiency in terms of mapping space, when adjoining hexagons can efficiently and completely fill a space without overlap. An overall comparison of its efficiency when compared with the Central Composite and Box–Behnken approaches is presented in Table 4.18 (Ferreira et al. 2004). For the values of k (variables), the Doehlert design is the most efficient.

Consider a modern environmental application employing the Doehlert matrix design. Vanloot et al. (2007) used a combination a factorial design and the Doehlert design matrix approach for the solid-phase extraction of residual aluminum coagulants in treated waters. The research intent was to determine factors affecting the extraction-elution steps of aluminum onto a modified commercial resin. This is highly important since this cation is known to be associated with various health problems including gastrointestinal damage, renal oestrodistrophy, and Alzheimer's disease (Flaten 2001, Kametani and Nagata 2006).

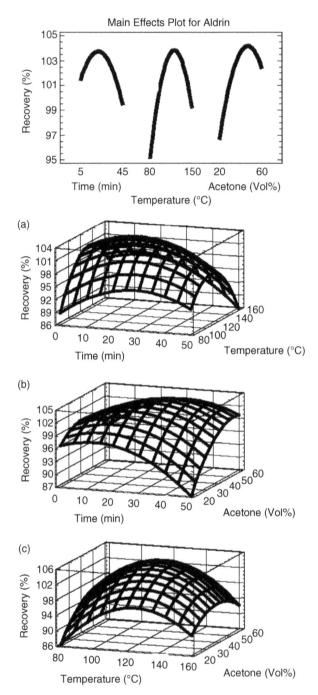

FIGURE 4.10 Main effects plot and estimated response surfaces for the central level of vol.% of acetone (a), extraction temperature (b), and extraction time (c) for the MAE of Aldrin. (From Gfrerer, M. and Lankmayr, E., *Analytica Chimica Acta*, 533, 203–211, 2005. With permission from Elsevier.)

TABLE 4.18

Comparison of Efficiencies of Doehlert Matrix Design (DM), Central Composite Design (CCD), and Box–Behnken Design (BBD) Approaches

Variables	Number of Coefficients (p)	Number of Experiments (f)						Efficiency (p/f)		
		CCD	DM	BBD		CCD		DM		BBD
2	6	9	7	—		0.67		0.86		—
3	10	15	13	13		0.67		0.77		0.77
4	15	25	21	25		0.60		0.71		0.60
5	21	43	31	41		0.49		0.68		0.61
6	28	77	43	61		0.36		0.65		0.46
7	36	143	57	85		0.25		0.63		0.42
8	45	273	73	113		0.16		0.62		0.40

Source: Adapted from Ferreira, S. L., et al., *Talanta*, 63, 1061–1067, 2004. With permission from Elsevier.

The matrix design for this study consisted of 43 distinct experiments and central points repeated three times to evaluate repeatability of measurements (Table 4.19). From these results, coefficient estimates were calculated using multilinear regression for aluminum extraction yield (Y_1), aluminum elution (Y_2), and aluminum recovery rates (Y_3). The former depends only on X_1, X_2, and X_3. After validation, the investigators utilized the regression equations to represent the responses in the whole experimental domain by RSM. For purposes of brevity, the investigators presented results of Y_3 to identify the step of solid-phase extraction that had the most influence on aluminum recovery rates. A closer look at the contour plots revealed that for an optimum Y_3 to 95%, HCl volume had to be at least 1.77 mL, with an elution flow rate below 3.9 mL/min (Figure 4.11a). The second plot (Figure 4.11b) confirms that 95% yield was obtained when the elution flow rate was below 3.9 mL/min with HCl concentration below 0.40 mol/L.

4.8 MIXTURE DESIGNS

In *mixture design* experiments, the independent factors are proportions of different components of a blend or mixture (e.g., eluents in liquid chromatography) and are often measured by their portions, which sum up to 100% or normalize to 1, that is,

$$\sum_{i=1}^{n} X_i = 1 \quad \text{for } x_i \geq 0 \tag{4.9}$$

Unlike the screening designs covered in Section 4.5, mixture designs are not orthogonal, and it is impossible to vary one factor independently of all others present. As assumed for a mixture, when the proportion of one ingredient changes, the proportion of one or more of the other ingredients must therefore also change to compensate. The proportions sum to one and thus present an interesting geometry. This in return, has significant effects on the factor space, the design properties, and the overall interpretation of the results. The feasible region for the response in a mixture design takes the form of a *simplex*. A simplex is $n + 1$ geometric figures in an n-dimensional space, where each vertex corresponds to a set of chosen experimental conditions.

The shapes of the simplex in a one-, a two-, and a three-variable space are a line, a triangle, or a tetrahedron. Such designs are utilized frequently in environmental analysis, especially with regard to assessing mixture interactions (e.g., PAH mixtures), and instrumental optimization.

As shown in Equation 4.9, components are subject to the constraint that they must equal to the sum of one. In this case, standard mixture designs for fitting standard models such as *simplex centroid* and *simplex lattice* are employed. When mixtures are subject to additional constraints, constrained mixture designs are then appropriate. Like the factorial experiments discussed earlier, mixture experimental errors are independent and identically distributed with zero mean and common variance. In addition, the true response surface is considered continuous over the region being studied. Overall, the measured response is assumed to depend only on

TABLE 4.19

Doehlert Design and Experimental Matrix for Solid-Phase Extraction of Aluminum Coagulants in Treated Waters[a]

Experiment	X_1 (mL)	X_2 (mL/min)	X_3 (µg/L)	X_4 (mL)	X_5 (mL/min)	X_6 (mol/L)	Y_1 (%)	Y_2 (%)	Y_3 (%)
1	5.00	4.75	110.00	2.60	4.75	0.30	88.20	86.60	76.40
2	0.50	4.75	110.00	2.60	4.75	0.30	89.30	93.00	83.00
3	3.88	8.45	110.30	2.60	4.75	0.30	79.80	87.60	69.90
4	1.63	1.05	110.30	2.60	4.75	0.30	96.50	91.50	88.30
5	3.88	1.05	110.00	2.60	4.75	0.30	96.00	89.20	85.60
6	1.63	8.45	110.00	2.60	4.75	0.30	83.70	93.40	78.10
7	3.88	6.00	185.00	2.60	4.75	0.30	83.70	89.10	74.60
8	1.63	3.50	35.00	2.60	4.75	0.30	86.90	87.50	76.10
9	3.88	3.50	35.00	2.60	4.75	0.30	91.50	89.50	81.90
10	2.75	7.20	35.00	2.60	4.75	0.30	82.30	92.50	76.10
11	1.63	6.00	185.00	2.60	4.75	0.30	85.50	91.50	78.20
12	2.75	2.30	185.00	2.60	4.75	0.30	93.40	88.50	82.70
13	3.88	6.00	128.75	4.50	4.75	0.30	85.90	90.50	77.70
14	1.63	3.50	91.25	0.70	4.75	0.30	87.90	86.40	75.90
15	3.88	3.50	91.25	0.70	4.75	0.30	88.40	83.70	74.00
16	2.75	7.20	91.25	0.70	4.75	0.30	81.70	85.90	70.20
17	2.75	4.75	166.25	0.70	4.75	0.30	87.60	84.90	74.30
18	1.63	6.00	128.75	4.50	4.75	0.30	86.50	93.80	81.20
19	2.75	2.30	128.75	4.50	4.75	0.30	92.20	92.80	85.60
20	2.75	4.75	53.75	4.50	4.75	0.30	88.20	92.80	81.80
21	3.88	6.00	128.75	2.98	8.10	0.30	86.70	74.50	64.60
22	1.63	3.50	91.25	2.22	1.40	0.30	86.50	97.40	84.30
23	3.88	3.50	91.25	2.22	1.40	0.30	88.40	94.90	83.80
24	2.75	7.20	91.25	2.22	1.40	0.30	81.70	96.10	78.50

continued

TABLE 4.19 (continued)

Experiment	X_1 (mL)	X_2 (mL/min)	X_3 (µg/L)	X_4 (mL)	X_5 (mL/min)	X_6 (mol/L)	Y_1 (%)	Y_2 (%)	Y_3 (%)
25	2.75	4.75	166.25	2.22	1.40	0.30	87.60	94.80	83.00
26	2.75	4.75	110.00	4.12	1.40	0.30	86.30	97.40	84.00
27	1.63	6.00	128.75	2.98	8.10	0.30	86.50	78.00	67.50
28	2.75	2.30	128.75	2.98	8.10	0.30	93.00	76.00	70.70
29	2.75	4.75	53.75	2.98	8.10	0.30	86.20	78.30	67.50
30	2.75	4.75	110.00	1.08	8.10	0.30	86.30	77.30	66.70
31	3.88	6.00	128.75	2.98	5.3	0.45	86.70	85.20	73.90
32	1.63	3.50	91.25	2.22	4.2	0.15	87.90	95.90	84.30
33	3.88	3.50	91.25	2.22	4.2	0.15	88.40	93.20	82.40
34	2.75	7.20	91.25	2.22	4.2	0.15	82.90	94.70	78.50
35	2.75	4.75	166.25	2.22	4.2	0.15	86.10	92.10	79.40
36	2.75	4.75	110.00	4.12	4.2	0.15	85.20	95.10	81.00
37	2.75	4.75	110.00	2.60	7.55	0.15	86.30	75.80	65.40
38	1.63	6.00	128.75	2.98	5.3	0.45	85.60	88.00	75.40
39	2.75	2.30	128.75	2.98	5.3	0.45	92.20	86.80	80.00
40	2.75	4.75	53.75	2.98	5.3	0.45	86.20	90.50	78.00
41	2.75	4.75	110.00	1.08	5.3	0.45	86.30	77.30	66.70
42	2.75	4.75	110.00	2.60	1.95	0.45	85.20	96.80	82.50
43	2.75	4.75	110.00	2.60	4.75	0.45	86.30	91.00	78.50
44	2.75	4.75	110.00	2.60	4.75	0.30	86.30	91.00	78.50
45	2.75	4.75	110.00	2.60	4.75	0.30	85.20	90.90	77.50

[a] X_1 = flow-through sample volume; X_2 = sample percolation rate; X_3 = sample metal concentration; X_4 = volume of HCl; X_5 = HCl flow rate; X_6 = [HCl]; Y_1 = Al^{3+} extraction rate; Y_2 = Al^{3+} elution rate; Y_3 = Al^{3+} recovery rates.

Source: Modified from Vanloot, P., et al., *Talanta*, 73, 237–245, 2007. With permission from Elsevier.

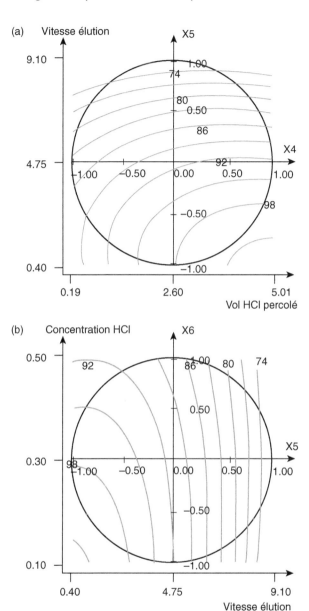

FIGURE 4.11 Study of elution parameters: (a) variation of the response Y_3 in the plane: X_4 (HCl volume), X_5 (elution flow-rate), and fixed factors, flow-through sample volume = 2.75 mL, sample percolation rate = 4.75 mL min^{-1}, sample metal concentration = 110 µg L^{-1}, and HCl concentration = 0.30 mol L^{-1}, (b) variation of the response Y_3 in the plane: X_5 (elution flow-rate), X_6 (HCl concentration), and fixed factors, flow-through sample volume = 2.75 mL, sample percolation rate = 4.75 mL min^{-1}, sample metal concentration = 110 µg L^{-1}, and HCl volume = 2.60 mL. (From Vanloot, P., et al., *Talanta*, 73, 237–245, 2007. With permission from Elsevier.)

the relative proportions of the components in the mixture and not on the amount. The following closer examination of commonly used techniques will facilitate understanding.

4.8.1 SIMPLEX CENTROID DESIGNS

Such designs are of degree k with n factors composed of proportionate mixture runs with the following combinations:

1. Single-factor combinations.
2. Two-factor combinations at all levels.
3. Three-factor combinations at equal levels.
4. Center point run with equal amounts of mixture.

Consider a hypothetical study of the fully resolved separation of a mixture of two isomeric compounds, A and B, by reverse phase high performance liquid chromatography (HPLC) utilizing the simplex centroid design. The resolution of isomeric compounds is a classic problem in the chemical sciences because the selectivity factor for compounds with identical masses, identical functionality, and similar structures is poor. Failure to identify isomeric compounds specifically, compromises one's ability to calibrate an instrument with standards, and consequently, co-eluting isomeric compounds cannot be quantified (Gonzales et al. 2007). One approach to manage the resolution of such compounds is to adjust the mobile phase mixture. There is a strong dependence of the retention time on the mobile phase composition, and the retention parameter may be easily altered by the variation of solvent polarity. The mergence of the two parameters (resolution and retention time of the second peak, compound B) can be used to transform a dimensionless assessment value of D (see Section 4.9 for full explanation). This value is then used to determine optimization performance.

TABLE 4.20

Simplex Design with Three Factors (Methanol, Acetonitrile, and Methanol/Acetonitrile/Water)

Experimental Runs	Methanol	Acetonitrile	Methanol/Acetonitrile/Water	D
1	1	0	0	0.654
2	0	1	0	0.192
3	0	0	1	0.045
4	1/2	1/2	0	0.455
5	1/2	0	1/2	0.384
6	0	1/2	1/2	0.145
7	1/3	1/3	1/3	0.332

The simplex design with three factors [methanol (%), acetonitrile (%), and methanol/acetonitrile/water (%)] is presented in Table 4.20. The first three runs are for each single-factor combination, the second set for each combination of two factors, and the last run, for the center point. A two-dimensional representation in the form of a ternary plot is presented in Figure 4.12a. Such a plot shows how close to a proportion of 1 a given factor is, by how close it is to the vertex of that factor in the triangle.

Just as with previous designs, a fitted model can be obtained taking into account that all factors add up to a constant, and thus a traditional full linear model will not be fully estimable. A response surface model is generated with the following characteristics:

1. Suppression of the intercept.
2. Inclusion of all linear main-effect terms.
3. Inclusion of all the cross terms (e.g., $X_1 * X_2$).
4. Exclusion of all the square terms.

In this type of reduced model, termed a *Scheffé model* (one excluding an intercept) the coefficients on the linear terms are the fitted responses at the extreme points where the mixture consists of a single factor. The coefficients on the cross terms indicate the curvature across each edge of the factor space. As in the models presented earlier, final analysis can include whole model tests, ANOVA reports, and response-surface reports. Main effects and significant interactions in a mixture may be estimated through appropriate simplex centroid designs. The parameter estimates for this hypothetical study are presented in Table 4.21. The fitted quadratic model for the mixture is:

$$y = 0.654x_1 + 0.192x_2 + 0.045x_3 + 0.121x_1x_2 + 0.129x_1x_3 + 0.097x_2x_3 \quad (4.10)$$

A ternary plot can be generated with contours added via the model prediction formula (Figure 4.12b).

4.8.2 SIMPLEX LATTICE DESIGNS

A *simplex lattice* is a design of the set of all combinations where the factors' values are i/m, where i is an integer from 0 to m such that the sum of the factors is 1 or a total of

$$N = (k + m - 1)!/[(k - 1)!m!] \quad (4.11)$$

experiments where there are k factors (Brereton 2003). The lattice spans the mixture space more evenly by creating a triangular grid of experimental runs (Figure 4.13). Presented is a three-factor, three-level {3, 3} lattice design with the setup highlighted in Table 4.22. Using Equation 4.11, there are 5!/(2!3!) = 10 experiments in total. Unlike the centroid design, it has no noticeable centroid run. A summary of various simplex lattice designs with total experimental runs is presented in Table 4.23.

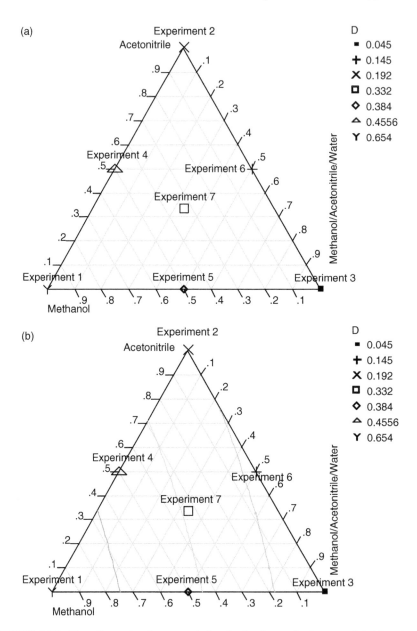

FIGURE 4.12 Two-dimensional representation in the form of a ternary plot (a) and with contours added via the model prediction formula (b).

TABLE 4.21
Parameter Estimates for the Hypothetical Simplex Study

Term	Estimate	Standard Error	t-ratio	Prob >|t|
Methanol	0.65445	0.00515	127.07	0.0050
Acetonitrile	0.19245	0.00515	37.37	0.0170
Methanol/acetonitrile/water	0.04545	0.00515	8.82	0.0718
Methanol * acetonitrile	0.1214	0.023675	5.13	0.1226
Methanol * methanol/acetonitrile/water	0.129	0.023675	5.45	0.1156
Acetonitrile * methanol/acetonitrile/water	0.097	0.023675	4.10	0.1524

What about mixtures containing multicomponents, say a mixture of PAHs, where the toxicity or mutagenicty must be assessed? Fractional simplex designs can be utilized as a means of screening for significant interactions present within a mixture without testing all possible binary combinations. Such designs consist of design points containing either one or p nonzero factor-level settings, where $p \geq 3$, such that all possible binary combinations occur only once within the design array (McConkey et al. 2000).

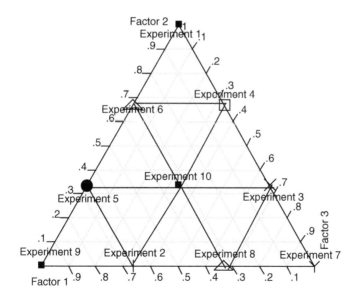

FIGURE 4.13   Triangular grid of experimental runs in the mixture space lattice.

TABLE 4.22
A Three-Factor, Three-Level {3, 3} Lattice Design Matrix

Experimental Runs	Factor 1	Factor 2	Factor 3
1	0	0	1
2	0	0.33	0.66
3	0	0.66	0.33
4	0	1	0
5	0.33	0	0.66
6	0.33	0.33	0.33
7	0.33	0.66	0
8	0.66	0	0.33
9	0.66	0.33	0
10	1	0	0

4.8.3 SPLIT-PLOT MIXTURE DESIGNS

The benefits of using a simplex centroid design for chromatographic optimization is highlighted in Section 4.8.1. What if analyses required a two-mixture model for optimization, one for the chromatographic mobile phase and the other for the extraction solution? The use of the recently developed *split-plot* mixture designs can be beneficial for optimization under these conditions and can help describe the interactions between the two mixtures. A Scheffé polynomial is used to describe each of the two mixtures with the multiplication of the two polynomials resulting in a model in which all possible interactions between mobile phase and extraction solution proportions are represented by cross terms (Borges et al. 2007).

Borges et al. (2007) utilized a *split-plot* design for the fingerprint optimization of chromatographic mobile phases and extraction solutions for *Camellia sinensis,*

TABLE 4.23
A Simplex Lattice Design Structure with the Total Number of Experimental Runs Possible

Factors (k)	Two Interactions	Three Interactions	Four Interactions	Five Interactions	Six Interactions
2	3 runs				
3	6 runs	10 runs			
4	10 runs	20 runs	35 runs		
5	15 runs	35 runs	70 runs	126 runs	
6	21 runs	56 runs	126 runs	252 runs	462 runs

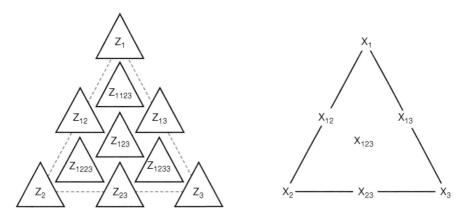

FIGURE 4.14 Experimental matrix for a split-plot design for the fingerprint optimization of chromatographic mobile phases and extraction solutions for *Camellia sinensis*. This design includes triangles representing the main-plots (extraction mixtures) for a simplex centroid design and a sub-plot representation (mobile phase mixture) of a simplex centroid design for each whole-plot. (From Borges, C. N., et al., *Analytica Chimica Acta*, 595, 28–37, 2007. With permission from Elsevier.)

a Chinese tea plant. The experimental design used is shown in Figure 4.14 and includes triangles representing the main plots (extraction mixtures) for a simplex centroid design and a subplot representation (mobile phase mixture) of a simplex centroid design for each whole plot. The x_{ijk} and z_{ijk} are Scheffé notations for mixtures. The seven simplex centroid triangles for the extraction solutions, each having triangles containing seven mobile-phase solutions, were used to evaluate a 49-term model obtained by multiplying cubic models for the mobile phase and the extraction solution mixtures.

The results for the experiments at the axial points were used to validate the mixture model calculated from the results of the simplex centroid extraction mixtures, and to calculate the number of peaks observed with the chromatographic detector at 210 nm. Detailed discussions on the cubic models used can be found within the main text of the study (Borges et al. 2007). Figure 4.15 shows the number of peaks observed with the detector at 210 nm for the whole-plot extraction mixture and subplot mobile phase mixture experimental design. The response surface examination is presented in Figure 4.16 and is based on model calculations. The x_1, x_2, and x_3 symbols represent the proportion of methanol, acetonitrile, and methanol-acetonitrile-water (MAW) 15%, 15%, 70% for the mobile phase and z_1, z_2, and z_3 for the ethyl acetate, ethanol, and dichloromethane proportions of the extracting solutions.

4.8.4 SIMPLEX OPTIMIZATION

The basic *fixed-size simplex* algorithm begins with the initial trials equal to the number of control factors plus one. As mentioned, the shapes of the simplex in a one-, two-, and three-variable space are a line, triangle, or tetrahedron. A geometric interpretation

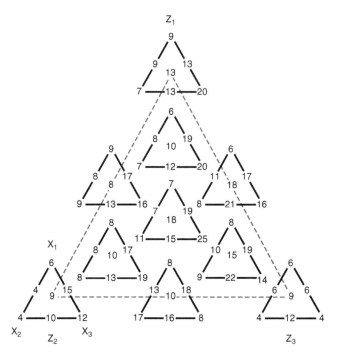

FIGURE 4.15 Number of peaks observed with the detector at 210 nm for a whole-plot extraction mixture and subplot mobile phase mixture experimental design. (From Borges, C. N., et al., *Analytica Chimica Acta*, 595, 28–37, 2007. With permission from Elsevier.)

is difficult when considering additional factors, but the basic steps (Brereton 2003) outlined here can handle the search for optimum conditions:

1. Define the number of factors (k).
2. Perform $k + 1$ experiments on the vertices of the simplex in the factor space.
3. Rank the response from 1 (worst) to $k + 1$ (best) over each of the initial conditions.
4. Establish new conditions for the next experiment using

$$x_{new} = c + c - x_1 \qquad (4.12)$$

where c is the centroid of the responses 2 to $k + 1$. Note that this excludes the worst response. Also keep the points x_{new} and the kth ($=2$) best responses from the previous simplex (results in $k + 1$ new responses).
5. Repeat as in steps 2 and 3 unless a response gained is worse than the remaining k ($=2$) conditions. If so, return to the previous conditions and calculate

$$x_{new} = c + c - x_2 \qquad (4.13)$$

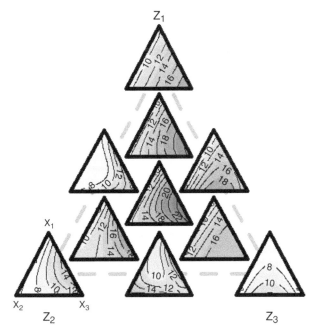

FIGURE 4.16 Response surface-type plot based on model calculations. (From Borges, C. N., et al., *Analytica Chimica Acta*, 595, 28–37, 2007. With permission from Elsevier.)

where c is the centroid of the responses 1 and 3 to $k + 1$. Note that this excludes the second-worst response.

6. Check for proper convergence. Note that if the same conditions reappear the analysis should be stopped.

Figure 4.17 shows an example of a fixed-size simplex optimization procedure. Shown are the changes in levels for two control factors. There are two essential rules to be kept when applying the simplex algorithm for optimization studies. First, reject the least favorable response value in the current simplex (after the initial trials that form the initial simplex). Second, never return to the control factor levels that have just been rejected. Additionally, there is a need to reevaluate the simplex for a specified number of steps and to avoid calculated trials outside the effective boundaries of the control factors.

The *modified simplex* approach attempts to correct for the weakness of the fixed-size, which is a dependence on initial step size (Brereton 2003). In this approach, the step width is changed by expansion and contraction of the reflected vertices (Figure 4.18). Two essential rules make up the modified approach. First, expansion is performed in a direction of more favorable conditions. Second, contraction occurs if a move is taken in a direction of less favorable conditions. The advantages of the modified approach include simplicity, speed, and good convergence properties (Otto 1999).

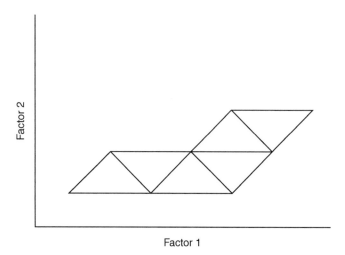

FIGURE 4.17   An example of a fixed-size simplex optimization procedure. Shown are the changes in levels for two control factors.

Based on these descriptions, can you name the advantages of the simplex approach over other optimization methods? Disadvantages? One advantage lies in the number of experiments performed. Here, only a single additional experiment is needed at each stage of the optimization process across the response area, while a factorial design requires at least 2^{k-1} trials. If the simplex size is too small, a large number of steps is required to reach the optimum point effectively. This problem, however, has been reduced by using a variable-size simplex approach.

To aid overall understanding, consider a study by Filipe-Sotelo et al. (2005) who utilized a simplex method in the optimization of molybdenum determination in solid environmental samples (e.g., coal fly ash slurries) by slurry extraction electrothermal atomic absorption spectroscopy (SE-ETAAS). Molybdenum (Mo) acts as an essential metal for animals, humans, and plants as well as a potential toxic in excess (Pyrzynska 2007). The investigators first employed a Plackett–Burman design (Section 4.5.3) to determine the factors that are important in the extraction of Mo from coal ash slurries. Results reveal that sample mass, ultrasonic power, HNO_3 concentration, and HCl concentration are the most significant. Multivariate simple

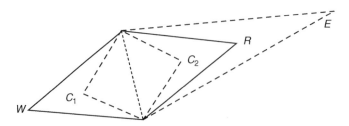

FIGURE 4.18   Modified simplex approach with the step width changed by expansion and contraction of the reflected vertices.

optimization of ultrasonic power, HNO_3, and HCl concentrations were performed due to interactions, with sample mass optimized individually.

Two simplex (regular and modified) were carried out keeping the coal fly ash mass constant (5 mg) and adopting experimental levels from the initial Plackett–Burman design with their results highlighted in Figure 4.19. For the purpose of interpretation, the first number of each vertex represents the order of the experimental point, with the response (% recovery) represented by the quantity after the slash. The first simplex (Figure 4.19a) evolved quickly with convergence after 10 movements. The final optimized conditions proved to be 15% HCl and 15% HNO_3 yielding ca. 80% extraction. The regular simplex (Figure 4.19b) needed a total of 17 movements reaching an improved response (ca. 90% recovery) at 14% HCl and 10% HNO_3. Both approaches ended in similar locations, thus suggesting that the experimental conditions were close to the *true* maximum in the chosen experimental region.

4.9 EXPANDED RESEARCH APPLICATION III—OPTIMIZED SEPARATION OF BENZO[A]PYRENE-QUINONE ISOMERS USING LIQUID CHROMATOGRAPHY-MASS SPECTROMETRY AND RESPONSE SURFACE METHODOLOGY

Chapter 1 discusses the ubiquitous presence in the environment, of *Polycyclic Aromatic Hydrocarbons* (PAHs), generated by fossil fuel and biomass combustion. These compounds have been linked to the development of chronic asthma and other respiratory problems in young children (Miller et al. 2004). Photodegradation is an important removal process for *Polycyclic Aromatic Compounds* (PACs) in ambient soil, air, and water (Nielson et al. 1983). Smog chamber studies have identified polar PAH photodegradation products, formed in the presence of various air pollutants such as SO_2, O_3, and NO_2 (Atkinson and Arey 1994, Finlayson-Pitts and Pitts 1997). Despite the impact of PAHs, and possibly their degradation products (benzo[a]pyrene-quinone isomers), on human health, further elucidation of the detailed mechanisms of the degradation processes are limited by the availability of sensitive, selective, and quantitative analytical techniques to characterize the quinones. We therefore concentrate on the work of Gonzales et al. (2007) who use response surface methodology in their development of a sensitive liquid chromatography-mass spectrometry (LC-MS) technique that specifically identify benzo[a]pyrene (BaP)-1,6/3,6-quinone peaks in complex environmental samples.

4.9.1 STATEMENT OF PROBLEM

Traditionally, investigators have relied on univariate approaches to optimize the resolution of analytes with low selectivity factors, including isomers. This approach is relatively simple and suitable for factors that are independent. However, univariate methods do not take interactive effects among factors into account, thus limiting the possibility of fully optimized separation and resolution. For example, HPLC has been used previously to analyze benzo[a]pyrene-quinones, known photodegradation products of BaP (Eisenberg et al. 1984), in environmental and laboratory samples (Koeber et al. 1999, Letzel et al. 2001). Koeber et al. (1999) discuss the detection of BaP-1,6, -3,6, and -6,

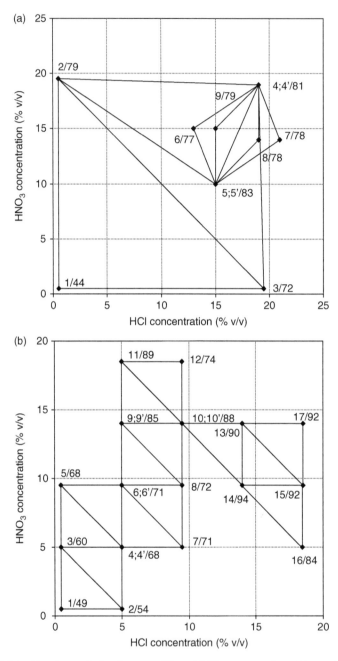

FIGURE 4.19 Simplex optimization of HNO$_3$ and HCl concentrations: (a) massive contraction simplex and (b) regular simplex. (From Filipe-Sotelo, M., et al., *Analytica Chimica Acta*, 553, 208–213, 2005. With permission from Elsevier.)

12-quinones extracted from real samples of air particulate matter. The data show three distinct peaks for these compounds but their poor resolutions, however, limit the application of this technique to more complex systems (e.g., biological, soil, water).

4.9.2 RESEARCH OBJECTIVES

To obtain a time-efficient resolution of complicated isomeric compounds using LC-MS with a combination of Derringer's desirability function and a Box–Behnken response surface design.

4.9.3 EXPERIMENTAL METHODS AND CALCULATIONS

As stated in Section 4.3, investigators must define the response and factors (and their levels) that will influence this response. Gonzales et al. (2007) considered two independent parameters, *resolution* and *retention time*, in their optimization approach for BaP-quinone separation. Resolution between the two consecutive quinone peaks was first calculated as

$$R = \frac{2(t_2 - t_1)}{w_1 + w_2} \tag{4.14}$$

where t and w denote the retention time and the corresponding baseline peak widths of the two consecutive peaks. The time corresponding to the peak intensity of the second compound eluding from the column was recorded as the retention time. The mergence of the two parameters (resolution and retention time) was accomplished by the use of Derringer's two-sided transformation (Derringer and Suich 1980), where each measured response was transformed into a dimensionless desirability (d). The values for desirability were set at zero for the undesirable level of quality, increasing to a more favorable desirability, to a maximum value of one. This approach is applicable when the response variable has both a minimum and a maximum constraint. They considered the transformations of both retention time and resolution by

$$d_i = \begin{cases} \left[\dfrac{Y_i - Y_{i*}}{c_i - Y_{i*}}\right]^s & Y_{i*} \leq Y_i \leq c_i \\[2ex] \left[\dfrac{Y_i - Y_i^*}{c_i - Y_i^*}\right]^t & c_i < Y_i \leq Y_i^* \\[2ex] 0 & Y_i < Y_{i*} \quad Y_i > Y_i^* \end{cases} \tag{4.15}$$

where Y_{i*} is the minimum acceptable value, and Y_i^* is the highest acceptable value of Y_i, the measured response. A value above the maximum acceptable value or below the minimum acceptable value would yield an unacceptable result of $d_i = 0$. In addition, c_i, represents the chosen target response as 20 minutes for the retention time transformation, and 2.0 as the optimum for resolution transformation. Exponents s and t represent the curvature from the minimum acceptable value to the target value, and the curvature from the target value to the maximum acceptable value, respectively. These in turn affect the curvature of the transformation graph and permit the investigators to weigh the relative importance of each target while producing a single overall desirability. Accordingly, measured values were transformed to a desirability that optimized peak resolution and retention time simultaneously. The individual desirabilities were then combined into a single assessment value of D using the geometric mean as follows

$$D = (d_{rn} \times d_{rt})^{1/2} \tag{4.16}$$

where d_{rn} and d_{rt} represent the transformations of resolution and retention time, respectively.

They next employed a Box–Behnken response surface design to locate the optimum flow rate, eluent composition, and column temperature conditions for separation by mapping the chromatographic response surface. How do you think the investigators selected the factors and levels needed? Factors and ranges considered in this study were based on previous univariate studies and chromatographic intuition. Eluent (the solvent phase used to move the *solute* through the column) composition (defined in this study as the volume of methanol with respect to total volume of solution) and column temperature have obvious effects on chromatographic *selectivity* (α) and were therefore chosen. Although eluent flow rate does not affect the selectivity factor, it plays a definite role on the analysis time, and thus, on the selected optimization criteria. Moreover, flow rate was chosen as a factor to determine whether, in fact, significant interactive effects could be observed even if its single effect was not statistically significant. This is a point made obvious from our study of Section 4.5.

Resolution refers to the ability of a column to separate chromatographic peaks.

Retention time is the time between injection and the appearance of the peak maximum.

Table 4.24 shows the three chromatographic factors and the levels selected in which experimental optimization, in terms of overall response (D), was performed. The generalized model used in this study has the quadratic form

---

TABLE 4.24

Experimental Factors and Levels Used in the Box–Behnken Design

Factor	Level (−)	Level (0)	Level (+)
Eluent composition	0.60	0.75	0.90
Flow rate (μL min^{-1})	500	800	1100
Column temperature (°C)	20	25	30

Source: Adapted from Gonzales, A., Foster, K. L., and Hanrahan, G., *Journal of Chromatography A*, 1167, 135–142, 2007. With permission from Elsevier.

$$Y = \beta_0 + \beta_1 X_1 + \beta_2 X_2 + \beta_3 X_3 + \beta_{12} X_1 X_2 + \beta_{13} X_1 X_3$$
$$+ \beta_{23} X_2 X_3 + \beta_{11} X_1^2 + \beta_{22} X_2^2 + \beta_{33} X_3^2 \tag{4.17}$$

which contains linear, and squared, terms for all factors, and the products of all pairs of factors. In this study, X_1, X_2, and X_3 terms correspond to flow rate, column temperature, and eluent composition as they relate to predicting D. In this equation β is the coefficient, akin to a regression coefficient and giving a measure of the rate of change in D per unit change in flow rate, column temperature, or eluent composition. Factor significance was calculated in ANOVA models that were estimated and run up to their first-order interaction terms. If the probability associated with the F-ratio is small, the model is considered a better statistical fit for the data than the response mean alone.

4.9.4 RESULTS AND INTERPRETATION

The experimental design matrix generated for the Box–Behnken study is shown in Table 4.25. Figure 4.20 presents representative chromatographs from experiments 1, 3, 9, and 10 described in Table 4.25, and depicted in Figure 4.20a, b, c, and d panels, respectively. These results illustrate operating conditions with various degrees of

TABLE 4.25
Box–Behnken Design Matrix with Mean Actual and Model Predicted Responses

Experiment	Eluent Composition	Flow Rate (μL min^{-1})	Column Temperature (°C)	Mean Actual Response (D) ($n = 3$)	Model Predicted Response (D) ($n = 3$)
1	0.90	1100	25	0.074	0.050
2	0.90	500	25	0.175	0.113
3	0.60	1100	25	0.035	0.096
4	0.60	500	25	0.195	0.219
5	0.90	800	30	0.111	0.066
6	0.90	800	20	0.163	0.304
7	0.60	800	30	0.501	0.370
8	0.60	800	20	0.099	0.152
9	0.75	1100	30	0.413	0.483
10	0.75	1100	20	0.642	0.534
11	0.75	500	30	0.510	0.618
12	0.75	500	20	0.656	0.587
13	0.75	800	25	0.460	0.457
14	0.75	800	25	0.469	0.457
15	0.75	800	25	0.443	0.457

Source: Adapted from Gonzales, A., Foster, K. L., and Hanrahan, G., *Journal of Chromatography A*, 1167, 135–142, 2007. With permission from Elsevier.

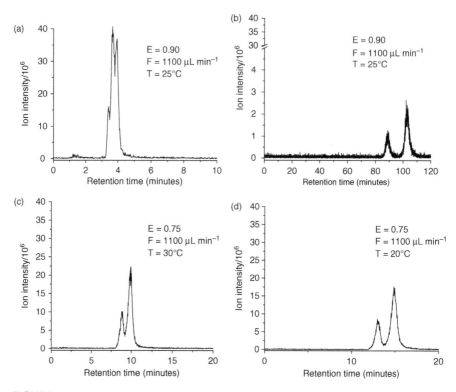

FIGURE 4.20 Representative chromatographs for the Box–Behnken design: (a) run 1, (b) run 3, (c) run 9, (d) run 10 with defined E, eluent composition; F, flow rate; and T, column temperature. (From Gonzales, A., Foster, K. L., and Hanrahan, G. *Journal of Chromatography A*, 1167, 135–142, 2007. With permission from Elsevier.)

optimization, quantified numerically as the mean actual Derringer desirability response (D) in column five of Table 4.25. It is important to note that the quantitative measure of optimization, D, corresponds with the qualitative assessment one can make by inspection. For example, although experiment three in Figure 4.20 yields peaks that are baseline resolved, the resulting retention time is over 100 minutes and consequently, the operating conditions used in Experiment 3 are the least desirable of the four experiments discussed. Included in Table 4.25 are the mean actual (experimental) and model-predicted responses, with the quality of fit expressed by the correlation coefficient and depicted visually by a whole-model leverage plot (Figure 4.21). The points on the plot are actual data coordinates with the horizontal line showing the sample mean of the response. The confidence curves (dashed lines) cross the horizontal line, and thus the test is considered significant at the 5% level. Overall, an r^2 value of 0.87 was obtained with a mean predicted response of 0.331.

Their examination of Prob > F from the Effect Test results (Table 4.26) revealed that flow rate and eluent composition had statistically significant single effects on the response, D. Remember that significance probabilities of ≤ 0.05 are often considered

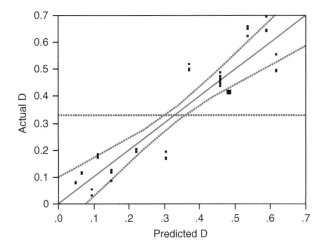

FIGURE 4.21 Whole-model leverage plot of actual vs. predicted responses. (From Gonzales, A., Foster, K. L., and Hanrahan, G., *Journal of Chromatography A*, 1167, 135–142, 2007. With permission from Elsevier.)

evidence that there is at least one significant regression factor in the model. Are there any other noticeable occurrences in Table 4.26? Interestingly, column temperature was not significant as an individual factor but displayed a strong interactive effect with eluent composition (<0.001). They postulated that BaP-quinones are more soluble in methanol-rich eluent than in eluent with less methanol and more water. Solubility is temperature dependent; hence the combined impact of eluent composition and temperature is predicted to be greater than that of either of these parameters alone. The coupled-interaction elucidated by this multivariate optimization technique could not be determined based on traditional univariate techniques alone.

TABLE 4.26
Effect Test Results for the Box–Behnken Design

Source	DOF	Sum of Squares	F-Ratio	Prob > F
Eluent composition	1	0.0344	4.868	0.0340
Flow rate	1	0.0525	7.3804	0.0102
Column temperature	1	0.0006	0.0857	0.7715
Eluent composition/flow rate	1	0.0026	0.3708	0.5465
Eluent composition/column temperature	1	0.1564	21.97	<0.001
Flow rate/column temperature	1	0.0051	0.7141	0.4038

Source: Adapted from Gonzales, A., Foster, K. L., and Hanrahan, G., *Journal of Chromatography A*, 1167, 135–142, 2007. With permission from Elsevier.

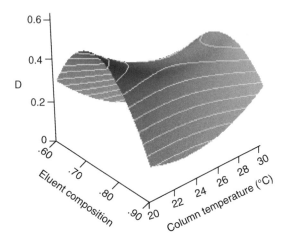

FIGURE 4.22   (See color insert following page 206.) Response surface image for the main interactive effect of eluent composition/column temperature at predicted critical values with flow rate held constant. (From Gonzales, A., Foster, K. L., and Hanrahan, G., *Journal of Chromatography A*, 1167, 135–142, 2007. With permission from Elsevier.)

Utilization of Equation 4.17 allowed the investigators to generate a response surface image (Figure 4.22) for the preceding main interaction. Here, they assessed how the predicted responses change with respect to changing these factors simultaneously, while keeping flow rate constant. The canonical curvature report (Table 4.27) shows the eigenstructure, which is useful in identifying the shape and orientation of the curvature and results from the eigenvalue decomposition of the matrix of second-order parameter estimates. See Appendix III for discussions on eigenvalues. The eigenvalues are negative if the response surface shape curves back from a maximum. The eigenvalues are positive if the surface shape curves up from a minimum. If the eigenvalues are mixed (as in this study), the surface is saddle shaped, curving up in one direction and down in another direction.

A post-hoc review of the model revealed optimum critical values of: eluent composition = 0.74, flow rate = 1100 µL min^{-1}; and column temperature = 22.0°C and a predicted D of 0.454. A series of three validation experiments using 5 µg mL^{-1}

TABLE 4.27

Canonical Curvature Results for the Response Surface Model in Gonzales et al. (2007)

	Eigen Vectors		
Eigenvalue	0.0498	−0.0438	−0.4361
Eluent composition	−0.11043	−0.01287	0.99380
Flow rate	−0.07131	0.99744	0.00499
Column temperature	0.99132	0.07032	0.11106

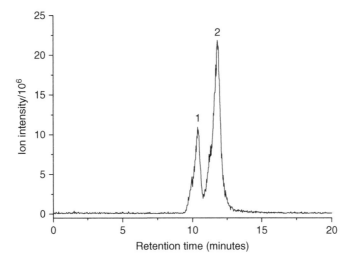

FIGURE 4.23 Chromatograph of BaP-1,6, and -3,6-quinones each at a concentration of $5\,\mu g\,mL^{-1}$ solvated in methylene chloride. Independent runs of each isomer were used to identify peaks 1 and 2 as BaP-1,6-quinone and BaP-3,6-quinone, respectively. (From Gonzales, A., Foster, K. L., and Hanrahan, G., *Journal of Chromatography A*, 1167, 135–142, 2007. With permission from Elsevier.)

standard solutions of benzo[a]pyrene-1,6 and -3,6-quinones were performed at the predicted critical values. A mean experimental D of 0.497 was obtained with an 8.9% discrepancy difference from the model predicted.

Under the optimum predicted critical values, a series of three seven-point calibration studies for both quinones were performed. Reproducibility for replicate injections of BaP-1,6-quinone standards ($0-12\ \mu g\ mL^{-1}$) is typically $<5.0\%$ R.S.D. ($n = 3$) with mean linear regression coefficients of $r^2 = 0.9919$. For BaP-3, 6-quinone standards ($0-12\ \mu g\ mL^{-1}$), reproducibility is typically $<5.0\%$ R.S.D. ($n = 3$) with good linear correlation ($r^2 = 0.9934$). A representative chromatograph obtained using the optimum critical values for a standard containing both isomers is contained in Figure 4.23. These results show that the optimal operating conditions yield a resolution of 0.98 within a 12.0 minute analysis time. The limit of detection (LOD) has been determined as the analyte concentration giving a signal equal to the blank signal, y_b, plus three times the standard deviation of the blank, s_b. Overall, their method achieves LODs of $0.604\ \mu g\ mL^{-1}$ and $0.270\ \mu g\ mL^{-1}$ for BaP-1, 6-quinone, and BaP-3,6-quinone, respectively.

Successful identification of three independent benzo[a]pyrene-quinone isomers produced in the irradiation of a benzo[a]pyrene standard was then shown using the optimized conditions predicted by the Box–Behnken model. Figure 4.24 presents the representative chromatograph of a $34.1\ \mu g\ mL^{-1}$ benzo[a]pyrene sample solvated in an oxygen-saturated methanol/methylene chloride solution irradiated for 10 minutes at room temperature. These data were obtained using the optimal flow rate, column temperature, and eluent composition identified by the Box–Behnken model mentioned earlier. These results show three new peaks that were not observed in a

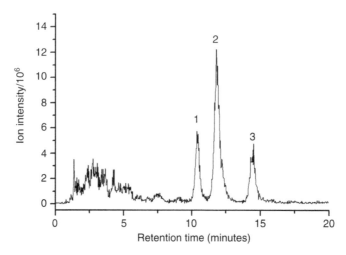

FIGURE 4.24   Chromatograph of 34 μg mL^{-1} BaP after 10 minutes of irradiation. (From Ferreira, S. L., et al., *Talanta*, 63, 1061–1067, 2004. With permission from Elsevier.)

34.1 μg mL^{-1} BaP sample analyzed prior to irradiation. In addition, the retention times of peaks 1 (10.38 minutes) and 2 (11.79 minutes), are in agreement with the observed retention times of BaP-1,6 and -3,6-quinone calibration standards presented in Figure 4.23.

4.9.5   Summary and Significance

In summary, the investigation presents significant progress in resolving complex isomeric compounds using LC-MS with a combination of Derringer's desirability function and response surface methodology. The predictive nature of a validated Box–Behnken design regarding the determination of the factors having the greatest influence on reaching an optimized response in a limited number of experiments is successfully presented. Successful identification of three independent benzo[a]pyrene-quinone isomers produced in the irradiation of a benzo[a]pyrene standard is shown utilizing optimized conditions. Such an approach is highly useful for investigators struggling with the coelusion of isomeric compounds in environmental samples and will likely be a major area of study for years to come.

4.10   CHAPTER SUMMARY

Chemometric experimental design and optimization techniques, and their applications in environmental analysis have been examined in detail. We have explored the concepts of factor screening, factorial analysis, response surface examination, system optimization, system robustness, and factor significance testing. The advantages of employing these techniques in modern environmental analysis, especially when compared with the use of traditional univariate methods, should be quite evident. In confirmation, an expanded environmental application is presented, incorporating many of the experimental design and optimization concepts covered in the chapter.

4.11 END OF CHAPTER PROBLEMS

1. Describe the differences between main and interactive effects in linear regression models.
2. In Section 4.1 you learned how to describe linear regression models for main and interactive effects. Produce a multiple linear regression model equation with three factors x_1, x_2, x_3 and one response, y. Include all possible terms and interactions in the model.
3. What impacts do randomization and blocking have on the ability of investigators to effectively plan and perform experimental design experiments?
4. The effects of four temperatures (50°C, 75°C, 40°C, and 60°C) upon the ability of a colorimetric method to determine effectively phosphate concentration in aqueous samples are to be investigated. Unfortunately, due to the time and amount of sample available, the investigator has to perform this in two batches with no more than eight runs per batch. Use the concepts of randomization and blocking to create a possible design for this study.
5. Draw a generalized graphical representation of the factor space of a 2^2 factorial design.
6. How many experiments are required in a 2^4 factorial design?
7. Name a primary disadvantage of utilizing a full or complete factorial design over a fractional factorial design.
8. Consider a 2^{5-1} fractional factorial design. (a) How many factors does this design have? (b) How many experimental runs does it require? (c) Construct a generic design matrix showing all coded levels with factors and the response represented by X and Y, respectively.
9. Compare the central composite, Box–Behnken and Doehlert matrix designs in terms of potential efficiency. How do the treatment of factors and levels differ in each approach?
10. What statistical tool is used to test the adequacy of response surface models?
11. Produce a design table for a three-factor, four-level simplex lattice design $\{3, 4\}$.

REFERENCES

Atkinson, R. and Arey, J. 1994. Atmospheric chemistry of gas-phase polycyclic aromatic hydrocarbons: formation of atmospheric mutagens. *Environmental Health Perspectives* 102: 117–126.

Borges, C. N., Bruns, R. E., Almeida, A. A., and Scarminio, I. S. 2007. Mixture-mixture design for the fingerprint optimization of chromatographic mobile phases and extraction solutions for *Camellia sinensis*. *Analytica Chimica Acta* 595: 28–37.

Bourguignon, B., Marcenac, F., Keller, H., de Aguiar, P. F., and Massart, D. L. 1993. Simultaneous optimization of pH and organic modifier content of the mobile phase for the separation of chlorophenols using a Doehlert design. *Journal of Chromatography* 628: 171–189.

Brereton, R. G. 2003. *Chemometrics: Data Analysis for the Laboratory and Chemical Plant.* West Sussex: Wiley.

Deming, S. N. and Morgan, S. L. 1993. *Experimental Design: a Chemometric Approach,* 2nd Ed. Chichester: Elsevier.

Derringer, G. and Suich, R. 1980. Simultaneous optimization of several response variables. *Journal of Quality Technology* 12: 214–219.

Doehlert, D. H. 1970. Uniform shell designs. *Applied Statistics* 19: 231–239.

Eisenberg, K., Taylor, K., and Murray, R. W. 1984. Production of singlet delta oxygen by atmospheric pollutants. *Carcinogenesis* 5: 1095–1096.

Ferreira, S. L. C., dos Santos, W. N. L., Quintella, C. M., Neto, B. B., and Bosque-Sendra, J. M. 2004. Doehlert matrix: a chemometric tool for analytical chemistry-review. *Talanta* 63: 1061–1067.

Filipe-Sotelo, M., Cal-Prieto, M. J., Carlosena, A., Andrade, J. M., Fernandez, E., and Prada, D. 2005. Multivariate optimization for molybdenum determination in environmental solid samples by slurry extraction-ETAAS. *Analytica Chimica Acta* 553: 208–213.

Finlayson-Pitts, B. J. and Pitts, J. N., Jr. 1997. Tropospheric air pollution: ozone, airborne toxics, polycyclic aromatic hydrocarbons, and particles. *Science* 276: 1045–1051.

Flaten, T. P. 2001. Aluminum as a risk factor in Alzheimer's disease, with emphasis on drinking water. *Brain Research Bulletin* 55: 187–196.

Gfrerer, M. and Lankmayr, E. 2005. Screening, optimization and validation of microwave-assisted extraction for the determination of persistent organochlorine pesticides. *Analytica Chimica Acta* 533: 203–211.

Giokas, D. L. and Vlessidis, A. G. 2007. Application of a novel chemometric approach to the determination of aqueous photolysis rates of organic compounds in natural waters. *Talanta* 71: 288–295.

Gonzales, A., Foster, K. L., and Hanrahan, G. 2007. Method development and validation for optimized separation of benzo[a]pyrene-quinone isomers using liquid chromatography-mass spectrometry and chemometric response surface methodology. *Journal of Chromatography A* 1167: 135–142.

Kametani, K. and Nagata, T. 2006. Quantitative elemental analysis on aluminum accumulation by HVTEM-EDX in liver tissues of mice orally administered with aluminum chloride. *Medical Molecular Morphology* 39: 97–105.

Koeber, R., Bayona, J. M., and Niessner, R. 1999. Determination of benzo[a]pyrene diones in air particulate matter with liquid chromatography mass spectrometry. *Environmental Science & Technology* 33: 1552–1558.

Lamas, J. P., Salgado-Petinal, C., García-Jares, C., Llompart, M., Cela, R., and Gomez, M. 2004. Solid-phase microextraction-gas chromatography-mass spectrometry for the analysis of selective serotonin reuptake inhibitors in environmental water. *Journal of Chromatography A* 1046: 241–247.

Letzel, T., Pöschl, U., Wissiack, R., Rosenberg, E., Grasserbauer, M., and Niessner, R. 2001. Phenyl-modified reversed-phase liquid chromatography coupled to atmospheric pressure chemical ionization mass spectrometry: a universal method for the analysis of partially oxidized aromatic hydrocarbons. *Analytical Chemistry* 73: 1634–1645.

López-García, I., Vizcaíno Martínez, M. J., and Hernández-Córdoba, M. 1991. Cold vapour atomic absorption method for the determination of mercury in iron(III) oxide and titanium oxide pigments using slurry sample introduction. *Journal of Analytical Atomic Spectrometry* 6: 627–630.

Madaeni, S. S. and Koocheki, S. 2006. Application of Taguchi method in optimization of wastewater treatment using spiral-wound reverse osmosis element. *Chemical Engineering Journal* 119: 37–44.

Massart, D. L., Vandeginste, B. G. M., Buydens, M. C., de Jong, S., Lewi, P. J., and Smeyers-Verbeke, J. 2003. *Handbook of Chemometrics and Qualimetrics*, Part A. Amsterdam: Elsevier.

McConkey, B. J., Mezey, P. G., Dixon, D. G., and Greenberg, B. M. 2000. Fractional simplex designs for interaction screening in complex mixtures. *Biometrics* 56: 824–832.

Miller, R. L., Garfinkel, R., Horton, M., Camann, D., Perera, F. P., Whyatt, R. M., and Kinney, P. L. 2004. Polycyclic aromatic hydrocarbons, environmental tobacco smoke, and respiratory symptoms in an inner-city birth cohort. *Chest* 126: 1071–1078.

Montgomery, D. C. 2005. *Design and Analysis of Experiments*, 6th Edition. New York: John Wiley & Sons.

Moreda-Piñeiro, J., López-Mahía, P., Muniategui-Lorenzo, S., Fernández-Fernández, E., and Prada-Rodríguez. 2002. Direct mercury determination in aqueous slurries of environmental and biological samples by cold vapour generation–electrothermal atomic absorption spectrometry. *Analytica Chimica Acta* 460: 111–122.

Nielson, T., Ramdahl, T., and Bjørseth, A. 1983. The fate of airborne polycyclic organic matter. *Environmental Health Perspectives* 47: 103–114.

Otto, M. 1999. *Chemometrics: Statistics and Computer Application in Analytical Chemistry.* Weinheim: Wiley-VCH.

Pyrzynska, K. 2007. Determination of molybdenum in environmental samples. *Analytica Chimica Acta* 590: 40–48.

Vanloot, P., Boudenne, J.-L., Vassalo, L., Sergent, M., and Coulomb, B. 2007. Experimental design approach for the solid-phase extraction of residual aluminum coagulants in treated waters. *Talanta* 73: 237–245.

Wu, C. F. J. and Hamada, M. 2000. *Experiments: Planning, Analysis, and Parameter Design Optimization.* New York: John Wiley & Sons.

5 Time Series Analysis

5.1 INTRODUCTION TO TIME SERIES ANALYSIS

The analysis of environmental data is inherently complex, with datasets often containing nonlinearities and temporal, spatial, and seasonal trends, as well as non-Gaussian distributions. *Time series analysis* attempts to address these concerns through the following operations:

1. the identification of the phenomenon represented by sequences of observations;
2. the characterization of patterns in a given dataset;
3. forecasting or prediction of future values of a given time series variable or set of variables.

A variety of time series models have been utilized for environmental analysis of variables where a record of observations at a particular location is associated with the progression of time. Selected time series examples with environmental applications are highlighted in Table 5.1. Detailed discussions on many of these techniques are presented in subsequent sections of this chapter.

Consider the environmental data (Table 5.2) and time series graph (Figure 5.1) showing total phosphorus (TP) measurements (μM) in a riverine system, for 31 consecutive days. The question we need to ask is: Are there any noticeable trends or fluctuations within the dataset? The *trend* component of time series analysis can reveal underlying directions (an upward or downward tendency) and rate of change in a particular time series analysis. Trend analysis uses both linear and nonlinear regression with time as the explanatory variable.

Now consider mean monthly TP measurement data over a one-year time frame (Table 5.3). Figure 5.2 shows the corresponding plot of the monthly data. The *seasonality* component reveals periodic fluctuations (e.g., monthly fluctuations) of a particular component. If seasonality is present, it must be incorporated into the time series model. As many environmental parameters (e.g., rainfall, discharge, nutrient concentrations) exhibit a sizeable degree of seasonality in their distribution, the methods of identifying the seasonality structure and the determination of a quantitative seasonal index are of considerable importance to investigators. The seasonal index represents the extent of seasonal influence for a particular portion of a given year and contains calculations involving the comparison of expected values for a portion with the overall mean. The seasonal index allows for seasonal adjustment in

TABLE 5.1
Selected Environmentally Related Time Series Studies

Parameter	Modeling Approach	Defining Characteristics/Results	Reference
Strong acidity (SA) Black smoke (BS)	Sequential approach to time series analysis (SATSA). This approach performed spectrum analysis through Fourier series decomposition.	Time series analysis of air pollution data identified periodic components of SA and BS in the Oporto, Portugal Metropolitan area.	Salcedo et al. (1999)
Chemical oxygen demand (COD) Chlorophyll-a (CHL-a) Nitrate Phosphate Total lead	Autoregressive integrated moving average (ARIMA) models.	Modeling via a series of ARIMA models revealed improvements of water quality due to the reduction of waste water emissions in the River Elbe, Germany.	Lehmann and Rode (2001)
Ozone (O_3)	Application of the Kolmogrov–Zurbenko (KZ) filter, the adaptive window Fourier transform (AWFT), the wavelet transform and the elliptic filter to spectrally decompose ozone time series.	The comparison of the four decomposition techniques revealed quantitatively similar results when used to extract frequency bands in the time series.	Hogrefe et al. (2003)
Rainfall	Fourier analysis utilizing the Hamming criteria to assess spectral concentration.	Distinct 12-month cycles were discovered in the time series analysis of rainwater sample volumes (winter and summer periods).	Astel et al. (2004)
Radon gas	Moving average smoothing and Fourier analysis.	Autocorrelation plots showed strong evidence of short range 24-hour periodicity. Cross-correlation analysis showed radon concentration correlations with rainfall, temperature, and to a lesser degree, tidal strength.	Groves-Kirkby et al. (2006)
Meteorological (rainfall, sunshine, relative humidity, barometric pressure and minimum/ maximum temperature) Air pollution (ozone and PM_{10})	Fourier analysis to recreate regular seasonal pattern. Logistic regression used to investigate all short-term parameter associations.	Time series analysis and logistic regression revealed no association between preterm births and recent exposure to ambient air pollution or changes in weather patterns.	Lee et al. (2008)

TABLE 5.2
TP Measurements over a 31-Day Time Period

Day	Daily TP Concentrations (μM)
1	3.98
2	4.45
3	3.88
4	5.00
5	3.88
6	4.43
7	4.55
8	3.81
9	3.75
10	3.71
11	4.57
12	3.97
13	5.00
14	3.84
15	3.76
16	3.75
17	5.66
18	4.04
19	4.50
20	4.56
21	4.99
22	4.03
23	3.99
24	3.76
25	3.69
26	3.62
27	4.23
28	4.00
29	3.46
30	3.38
31	2.40

forecasting and enables investigators to *deseasonalize* time series to reveal trends that might have been masked by the seasonality pattern (Zellner 2004).

The related *cyclical* component reveals any regular, nonseasonal fluctuations in a given time series. In either case, the pattern of observed time series data is identified and more formally described. Finally, an *irregular* component is what remains after the seasonal and trend components of a time series have been estimated and removed.

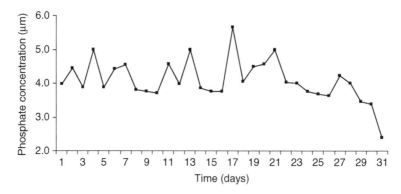

FIGURE 5.1   Time series plot of riverine TP measurements over 31 consecutive days.

5.2   SMOOTHING AND DIGITAL FILTERING

One basic assumption in any given time series component is the presence of random variation. Some form of data smoothing is thus necessary to filter out this noise in order to help reveal more significant underlying trends and seasonal components. The two main types of smoothing employed in time series analysis are *moving average smoothing* and *exponential smoothing* methods.

5.2.1   Moving Average Smoothing

Moving average smoothing is considered the easier of the two methods and replaces each element of the series by either the simple or weighted average of n surrounding elements where n is the width of the smoothing window (Brown 2004). Simple

TABLE 5.3
Mean Monthly TP Measurements

Month	Mean Monthly TP Concentrations (μM)
1	4.38
2	4.04
3	3.44
4	2.87
5	3.38
6	4.67
7	5.34
8	5.47
9	5.70
10	5.70
11	4.89
12	4.36

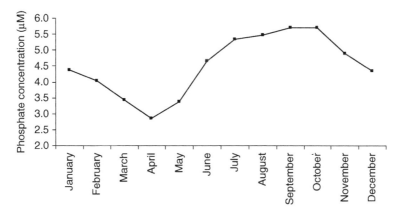

FIGURE 5.2 Time series plot of monthly TP measured over a year time frame.

(equally weighted) moving averages are frequently used to estimate the current level of a time series, with this value being projected as a *forecast* (see Section 5.3) for future observations. Ostensibly, forecasting uses a set of historical values (time series) to predict an outcome. The simple moving average is the average of the last *n* measurements or observations calculated as:

$$Y' = \frac{Y(t-1) + Y(t-2) + \cdots + Y(t-n)}{n} \tag{5.1}$$

where the one-period-ahead forecast $Y'(t)$ at $t - 1$ = the simple average of the last *n* measurements or observations. For example, a simple moving average command added to Figure 5.1 would result in the visual representation in Figure 5.3. Note that the average is centered at the period $t - (n + 1)/2$, implying that the estimate of the local mean will *lag* behind its true value by $(n + 1)/2$ periods. If one were to average the last five values in this times series of measurements, the forecasts will be roughly three periods late in responding to *turning points*.

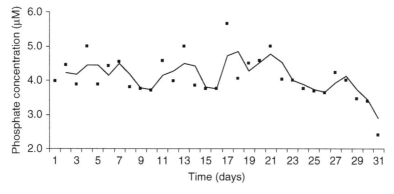

FIGURE 5.3 Time series plot of riverine TP measurements with an added simple moving average command.

A *turning point* is defined as a "local extremum" of a trend component of a time series.

5.2.2 EXPONENTIAL SMOOTHING

There are a variety of exponential smoothing methods employed in time series analysis. Such methods are weighted averages of past observations, with the weights decaying exponentially as the measurements or observations age (Athanasopoulos and Hyndman 2008). With the choice of a smoothing constant, a smoothed series can be generated by exponential smoothing. The smoothed value calculated for the final period can then be used as the forecast of future values. All exponential smoothing methods use this recurrence relation.

Consider the *single exponential smoothing* model, which works best for series exhibiting no marked seasonality or trend and is represented by

$$S_t = \alpha y_t + (1 - \alpha)S_{t-1} \qquad 0 < \alpha < 1 \tag{5.2}$$

where S_t is the corresponding smoothed value, α the smoothing weight, and y_t is the actual (noisy) value. Note that the higher the value assigned to α, the more the weight given to current values with the previous values suppressed. If a noticeable trend is present in the times series, this is the weaker choice, even when increasing the value of α to near 1.

The *double exponential smoothing* model improves on the single approach by the addition of a second equation with an added constant (β), which is chosen in conjunction with α. This model is represented by

$$S_t = \alpha y_t + (1 - \alpha)(S_{t-1} + T_{t-1}) \qquad 0 < \alpha < 1 \tag{5.3}$$

$$T_t = \beta(S_t - S_{t-1}) + (1 - \beta)T_{t-1} \qquad 0 < \gamma < 1 \tag{5.4}$$

where T_t is the corresponding contribution from the trend and β is the constant controlling the smoothing of the trend. Note that the first trend adjusts S_t for the trend of the previous period by adding it to the last smoothed value, S_{t-1}, to help eliminate potential lag. The primary role of the second equation is to update the trend.

If a given time series shows both trend and seasonality, the two models are no longer appropriate for such a situation. Instead, a third variant of exponential smoothing, the *Holt–Winters method*, which takes into account known seasonality (also termed periodicity) by introducing a third parameter, γ. The Holt–Winters model forecasts future values by estimating and extrapolating a linear trend, adjusting the data in each season by estimated seasonal indices and, based on weighted averages of all previous data values, using separate smoothing constants for each (Gardner 1985). There are both multiplicative and additive seasonal models with the four associated equations given by

$$S_t = \alpha \frac{y_t}{I_{t-L}} + (1 - \alpha)(S_{t-1} + T_{t-1}) \qquad \text{Overall smoothing} \tag{5.5}$$

$$T_t = \beta(S_t - S_{t-1}) + (1 - \beta)T_{t-1} \qquad \text{Trend smoothing} \qquad (5.6)$$

$$I_t = \gamma\frac{y_t}{S_t} = (1 - \gamma)I_{t-L} \qquad \text{Seasonal smoothing} \qquad (5.7)$$

$$F_{t+m} = (S_t + mT_t)I_{t-L+m} \qquad \text{Forecasting} \qquad (5.8)$$

where I is the seasonal index, F is the forecast at m periods along, γ is the seasonality constant and L is the number of periods in the seasonal pattern (e.g., $L = 4$ for quarterly).

EXAMPLE PROBLEM 5.1

1. Considering the following plot (Figure 5.4), discuss what type of exponential smoothing model that would be best suited for forecasting this time series.

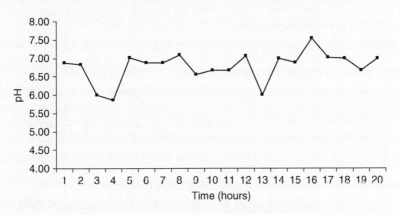

FIGURE 5.4   Example of pH time series plot of hourly measurement data.

Answer: Visual inspection of the plot does not reveal any noticeable trends or seasonal components in the time series. Therefore, the single exponential smoothing model would be the most appropriate.

5.3   TIMES SERIES AND FORECAST MODELING— A DETAILED LOOK

Time series modeling techniques are used to fit theoretical models to the time series and to make the fitted model predict (forecast) future values in the series. Each model discussed earlier has its own forecast plot showing the values that the models predict for the time series of interest. Greater understanding in choosing, building, and interpreting such plots are provided here.

5.3.1 MODEL CHOICE AND DIAGNOSTICS

The correct choice of smoothing model and model building are obvious considerations in time series analysis and forecasting. We have already discussed the process of choosing smoothing models depending on the type of time series data and any noticeable trends or seasonal component within the time series. After choosing the appropriate model, investigators must give logical thought to the model building process.

Consider the mean monthly (36 months) river flow (m³ s⁻¹) dataset presented in Table 5.4. Note that month 1 = January and yearly (12 month) progress is shown. Figure 5.5 displays a time series plot of the original data. From visual inspection, it appears that there are seasonal components to the time series. For example, rainy periods are likely to contribute to increased river flow during the winter and spring months. So what smoothing method would be appropriate in determining both trend and seasonality as well as providing the best model fit and forecasting capabilities? For purposes of comparison, let us consider the use of the double exponential smoothing and the Holt–Winters methods.

Figures 5.6a and b present the forecast plots generated from double exponential smoothing and the Holt–Winters method, respectively. As shown, these plots are divided into two regions by a vertical line. To the left of the separating lines, the one-step-ahead forecasts are shown overlaid with the input data points. To the right of the lines are the future values forecast (12 months) by the model and their 95% confidence intervals. Notice the inability of the double exponential smoothing method to forecast future monthly values effectively (Figure 5.6a). Compare this with the Holt–Winters method (Figure 5.6b), which shows future seasonal patterns similar to the original dataset. The reasons for this, and the differences between the models, are discussed in detail in the following Sections.

A closer look at the details of the time series in Figures 5.6a and 5.6b reveals many important aspects of the analysis. Each model generates statistical fit summaries (Table 5.5) by which to ascertain the adequacy of the model and its ability to forecast future values. As expected, the Holt–Winters method proved to be the more effective model to describe the time series, especially because of the seasonality components within the original dataset. Note the obvious differences in the statistical parameters between the models. If the model fits the series badly, the model error sum of squares will likely be larger than the total sum of squares, with the r^2 reported as negative. This is evident when examining the double exponential smoothing results. Note the differences in the *mean absolute error* (MAE) and the *mean absolute percentage error* (MAPE), which are calculated as

$$\text{MAE} = \frac{\sum_{i=1}^{n} |A_i - F_i|}{n} \tag{5.9}$$

TABLE 5.4
Mean Monthly River Flow (m³ s⁻¹) Values over a 36-Month Period

Month	Mean Monthly River Flow (m³ s⁻¹)
1	8.40
2	8.03
3	4.87
4	5.00
5	6.33
6	5.61
7	4.65
8	5.09
9	6.37
10	6.90
11	6.92
12	7.09
13	9.12
14	7.97
15	4.87
16	5.55
17	6.89
18	5.61
19	4.88
20	5.34
21	6.45
22	6.30
23	7.07
24	8.12
25	8.77
26	8.00
27	5.08
28	6.88
29	5.78
30	4.99
31	4.38
32	4.67
33	7.00
34	6.30
35	7.90
36	8.90

where MAE is the weighted average of the absolute errors, A_i the actual value, F_i the predicted value and n is the number of forecast periods. For the MAPE

$$\text{MAPE} = \frac{\sum_{t=1}^{n}\left(\left|\frac{A_i - F_i}{A_i}\right|\right) \times 100}{n} \qquad (5.10)$$

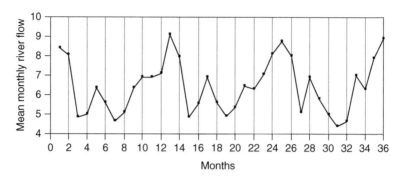

FIGURE 5.5   Time series plot of mean monthly river flow data.

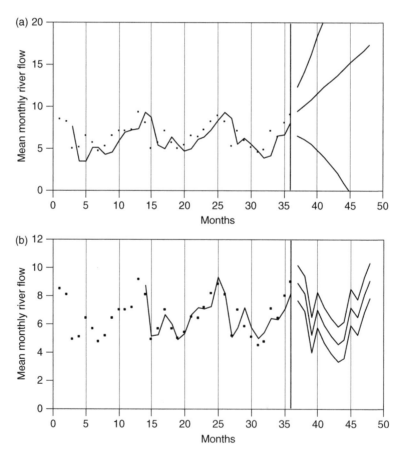

FIGURE 5.6   Time series plots of mean monthly river flow data: (a) incorporation of the double exponential smoothing method and (b) incorporation of the Holt–Winters method.

TABLE 5.5

Model Summary Statistics for Mean Monthly River Flow Times Series

Model/Parameter	Value
Double Exponential Smoothing	
Sum of squared errors	74.2
Variance estimate	2.25
Standard deviation	1.50
r^2	-0.34
MAPE	19.6
MAE	1.18
AIC	126
Holt–Winters Method	
Sum of squared errors	8.20
Variance estimate	0.41
Standard deviation	0.64
r^2	0.78
MAPE	7.94
MAE	0.50
AIC	50.3

Also, comparing models in terms of Akaike's information criteria (AIC), a measure of the goodness-of-fit (smaller values indicating better fit) of an estimated statistical model (Akaike 1974) is calculated as

$$\text{AIC} = 2k - 2\ln(L) \tag{5.11}$$

where k is the number of parameters in the statistical model and L is the likelihood function.

Plots of model residuals (actual values minus the one-step-ahead predicted values) based on the fitted models were also examined (Figure 5.7) to ensure that no obvious seasonal patterns remained in the time series. A good fit is characterized by points randomly distributed around 0, with no discernable patterns. In addition, the *auto-correlation* and *partial autocorrelation* of these residuals can be determined. Although more meaningful when using autoregressive integrated moving average (*ARIMA*) models (see Section 5.4), we can use them here to compare the two models.

Autocorrelation describes the correlation among all members of a time series with a given separation in time or lag. More specifically, autocorrelation occurs when residual error terms from the observations or measurements of the same variable at different times are in fact correlated. The autocorrelation for the kth lag is given by

$$r_k = \frac{c_k}{c_0} \qquad \text{where } c_k = \frac{1}{N} \sum_{t=k+1}^{N} (y_t - \bar{y})(y_{t-k} - \bar{y}) \tag{5.12}$$

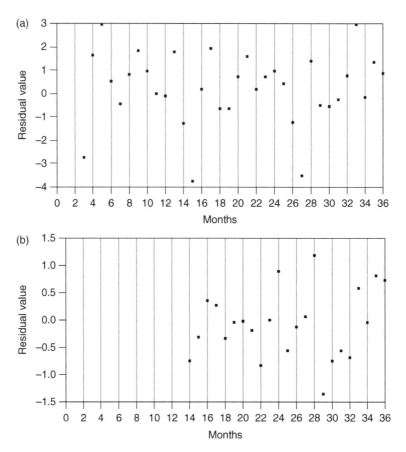

FIGURE 5.7 Plots of river flow residuals: (a) incorporation of the double exponential smoothing method and (b) incorporation of the Holt–Winters method.

In this equation, $\bar{y}$ is the mean of the N nonmissing points in the time series. The top portion, c_k, is similar to the covariance, but at a lag of k, and the bottom, c_0, is akin to the covariance at lag of 0. Autocorrelation plots can be used visually to show the set of values $\{r_k\}$ versus lag k in order to examine how the pattern of correlation varies with lag. In addition, the Ljung–Box Q- and p-values can be examined. The Q-statistic is used to test whether a group of autocorrelations is significantly different from zero, or to test the residuals from models that can be distinguished from *white noise* (Ljung and Box 1978). Both of these statistics are diagnostic measures of fit under the hypothesis that the model has accounted for all correlation up to the lag. The model is adequate if, for example, the correlation values of the lag sequence are less than 0.05 and the p-values are generally greater than 0.05. Carefully examine the diagnostic results presented in Tables 5.6 and 5.7. It is obvious that the Holt–Winters method provides a more appropriate fit based on these criteria. The representative autocorrelation column plots for the double exponential smoothing and the Holt–Winters method are shown in Figures 5.8a and b, respectively. Note that the

TABLE 5.6

Time Series Autocorrelation Diagnostics for the Use of the Double Brown Smoothing Model in the Mean Monthly River Flow Time Series

Lag	Autocorrelation	Ljung–Box Q	p-Value
0	1.0000	-	-
1	0.0823	0.2510	0.6164
2	−0.3344	4.5276	0.1040
3	0.0509	4.6298	0.2010
4	0.2602	7.3918	0.1166
5	−0.0816	7.6730	0.1752
6	−0.5068	18.900	0.0043
7	−0.0497	19.012	0.0081
8	0.1457	20.011	0.0103
9	−0.1343	20.894	0.0131
10	−0.3717	27.939	0.0018
11	0.1550	29.218	0.0021
12	0.5242	44.505	<0.001
13	−0.0150	44.518	<0.001
14	−0.1882	46.687	<0.001
15	0.1597	48.330	<0.001
16	0.1673	50.234	<0.001
17	−0.1255	51.367	<0.001
18	−0.2472	56.041	<0.001
19	0.0320	56.125	<0.001
20	0.0192	56.157	<0.001
21	−0.1972	59.820	<0.001
22	−0.1624	62.511	<0.001
23	0.0721	63.089	<0.001
24	0.1879	67.410	<0.001
25	−0.0708	68.092	<0.001

curved solid lines show twice the large-lag standard error (± 2 standard errors) and are computed as (JMP 2007)

$$\text{SE}_k = \sqrt{\frac{1}{N} \sum_{i=1}^{k-1} r_i^2} \qquad (5.13)$$

White noise corresponds to a series of uncorrelated random variables with zero mean and a given finite variance.

TABLE 5.7

Time Series Autocorrelation Diagnostics for the Use of the Holt–Winters Method in the Mean Monthly River Flow Time Series

Lag	Autocorrelation	Ljung–Box Q	p-Value
0	1.0000	—	—
1	0.0215	0.0121	0.9126
2	−0.1385	0.5373	0.7644
3	−0.0780	0.7121	0.8704
4	−0.0273	0.7346	0.9470
5	−0.1728	1.6884	0.8904
6	−0.4034	7.1934	0.3033
7	0.1599	8.1127	0.3228
8	0.1882	9.4697	0.3042
9	0.0547	9.5928	0.3844
10	−0.0899	9.9500	0.4449
11	0.1301	10.761	0.4635
12	0.1092	11.358	0.4962
13	−0.1843	13.338	0.4220
14	−0.1757	15.311	0.3572
15	0.0457	15.460	0.4188
16	0.0697	15.860	0.4627
17	0.0652	16.267	0.5049
18	0.0438	16.488	0.5585
19	0.0221	16.558	0.6197
20	0.0120	16.586	0.6796
21	−0.0898	18.904	0.5913
22	−0.0625	21.149	0.5115
23	0.0000	—	—
24	0.0000	—	—
25	0.0000	—	—

Partial autocorrelations are also useful in model diagnostics. This approach differs slightly from autocorrelations to the extent that the dependence on the intermediate elements within the lag is removed. If a lag of 1 is specified (no intermediate elements within the lag), the partial autocorrelation is considered equivalent to autocorrelation. Overall, it is believed that the partial autocorrelations provide a truer picture of dependencies for individual lags (Box and Jenkins 1970). Consider the relative diagnostics for the partial autocorrelations for the monthly river flow series (Tables 5.8 and 5.9) as well as the related partial autocorrelations plots for the double exponential smoothing and Holt–Winters method in Figures 5.9a and b, respectively. Note that the solid vertical curves represent ±2 standard errors for ~95% confidence limits, where the standard error is computed as (JMP 2007)

$$\mathrm{SE}_k = \frac{1}{\sqrt{n}} k \qquad (5.14)$$

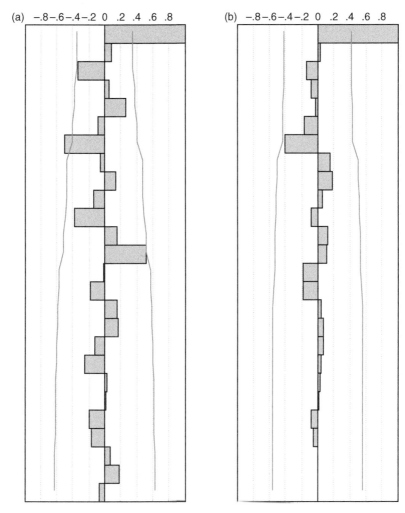

FIGURE 5.8 River flow autocorrelation plots for (a) the double exponential smoothing method and (b) the Holt–Winters method.

It is important to set the appropriate number of lags for the autocorrelation and partial autocorrelation plots when setting up analyses. This is the maximum number of periods between the points used in the computation of the correlations. As a general rule, the maximum number of lags is $n/4$ where n is the number of observations or measurements.

5.3.2 HARMONIC ANALYSIS TECHNIQUES

Think of *harmonic analysis* of a time series as a decomposition of the series. The term harmonic arises from the representation of functions or signals as the super-position of basic waves. Any periodic function that may thus result can be expressed

TABLE 5.8

Time Series Partial Autocorrelation Diagnostics for the Use of the Double Exponential Smoothing Model in the Mean Monthly River Flow Time Series

Lag	Partial Autocorrelation
0	1.0000
1	0.0823
2	−0.3435
3	0.1325
4	0.1413
5	−0.0906
6	−0.4333
7	−0.0254
8	−0.1539
9	−0.0998
10	−0.3334
11	0.1527
12	0.2325
13	0.0356
14	−0.0233
15	0.0170
16	−0.2826
17	0.0019
18	−0.0192
19	0.0376
20	−0.1680
21	0.0670
22	−0.0529
23	−0.1319
24	−0.1483
25	−0.0492

as the sum of a series of sines and cosines of varying amplitudes. Here, the coefficients of this decomposition are the discrete Fourier transform (DFT) of the series. A comprehensive review of Fourier analysis is beyond the scope of this book but detailed accounts can be found in dedicated sources (Blair 1995, Hirji 1997, Simpson and De Stefano 2004, Seeley 2006).

In general terms, the method for a Fourier series involving a function $f(x)$ can be represented by a series of sines and cosines as

$$f(x) = \frac{1}{2}a_0 + \sum_{n=1}^{\infty} a_n \cos(nx) + \sum_{n=1}^{\infty} b_n \sin(nx) \qquad (5.15)$$

TABLE 5.9
Time Series Partial Autocorrelation Diagnostics for the Use of the Holt–Winters Method in the Mean Monthly River Flow Time Series

Lag	Partial Autocorrelation
0	1.0000
1	0.0215
2	−0.1390
3	−0.0731
4	−0.0447
5	−0.1981
6	−0.4459
7	0.0856
8	0.0381
9	−0.0118
10	−0.1470
11	0.0351
12	−0.0222
13	−0.0448
14	−0.1184
15	−0.0046
16	−0.0670
17	0.1778
18	0.0532
19	−0.1186
20	−0.0750
21	0.0395
22	0.0036
23	0.0641
24	−0.0579
25	−0.0640

where

$$a_0 = \frac{1}{\pi} \int_{-\pi}^{\pi} f(x)\,dx$$

$$a_n = \frac{1}{\pi} \int_{-\pi}^{\pi} f(x)\cos(nx)\,dx$$

$$b_n = \frac{1}{\pi} \int_{-\pi}^{\pi} f(x)\sin(nx)\,dx \qquad n = 1, 2, 3, \ldots.$$

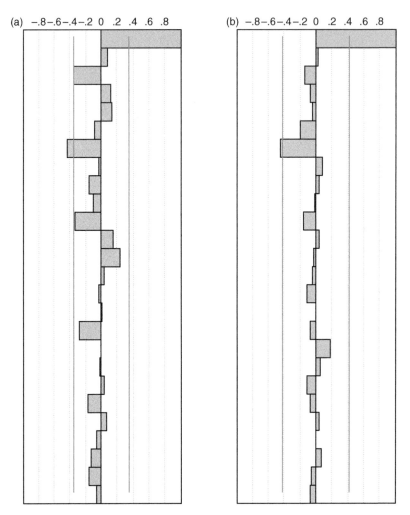

FIGURE 5.9 River flow autocorrelation plots for (a) the double exponential smoothing method and (b) the Holt–Winters method.

Fourier series can be further generalized to derive the *Fourier transform* (FT). The forward FT is utilized to map a given time series into a series of frequencies that compose the time series. It can be represented mathematically as

$$F(x) = \int_{-\infty}^{\infty} f(x)\,e^{-2\pi ikx}\,dk \qquad (5.16)$$

The inverse FT maps the series of frequencies back into the corresponding time series and is represented mathematically as

$$f(x) = \int_{-\infty}^{\infty} F(k)\,e^{2\pi ikx}\,dx \qquad (5.17)$$

Note that the two functions are inverses of each other and $e^{xi} = \cos(x) + i \sin(x)$. DFT is an operation to evaluate the FT of the sampled signal with a finite number of samples (N). The forward DFT is defined mathematically as

$$F_n = \sum_{k=0}^{N-1} f_k e^{-2\pi i n k/N}$$

(5.18)

The inverse DFT is defined as

$$f_k = \frac{1}{N} \sum_{n=0}^{N-1} F_n e^{-2\pi i k n/N}$$

(5.19)

The DFT is especially useful in environmental datasets to remove the high-frequency fluctuations (e.g., seasonal means) obtained from relevant measurements. Such datasets often have significant gaps of missing data, or are taken at irregular sampling intervals, or both. While the discrete transform can be applied to many complex series, it can take considerable time to compute, especially if large time series are of interest. The fast Fourier transform (FFT) is an efficient algorithm to compute a DFT. The most common FFT, developed by Cooley and Tukey, re-expresses the DFT of an arbitrary composite size in terms of a smaller DFT, recursively, in order to reduce the computation time to $0(n \log n)$ for highly-composite n (Cooley and Tukey 1965). The algorithm has also been shown to run approximately 200 times faster than a DFT on the same sample wave (Takahashi and Kanada 2000). Do note, however, that the FFT is limited in that it is most efficient when the time series has data sampled at equal intervals with no missing values (Dilmaghani et al. 2007).

As discussed, time series of environmental measurements typically display periodic or quasiperiodic behavior on a wide range of time scales. Another approach in gaining a better understanding of these processes is through a technique called spectral analysis. This approach transforms variation in time into an auto-spectrum that provides a representation of the amount variance of the time series as a function of frequency (Kay and Marple 1981). Mathematically, it can be defined as the FT of the autocorrelation sequence of a given time series or parametrically, via an approximation of the signal by ARIMA models (Section 5.4).

Prior to spectral analysis, the time series should be plotted to identify the presence of trends or seasonal components. Reconsider our study of mean monthly river flow data presented in Section 5.3.1 where the original time series plot is shown in Figure 5.5. The corresponding spectral density plot (*periodogram*) of this data is presented in Figure 5.10. As shown during smoothing model experiments, the presence of seasonal components was evident, although not easily seen in the periodogram (Figure 5.10). Note, however, the presence of strong spike-like noise and subsequent harmonics, which indicates the periodic character of the data. The predominance of frequency variability (frequencies <1) in the periodogram is also indicative of periodic variability in the time series (Chaloupka 2001).

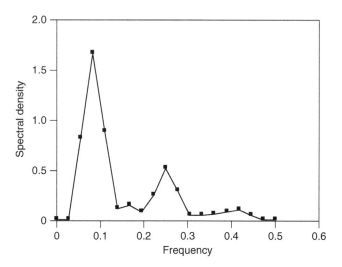

FIGURE 5.10   Spectral density plot of the original time series plot shown in Figure 5.5.

The Fisher's kappa statistical test can be used to test H_0 and whether the river flow series are drawn from a normal distribution with variance of 1. The probability of observing a larger kappa if H_0 is true is given by (JMP 2007)

$$(k > \kappa) = 1 - \sum_{i=0}^{q} (-1)^j \binom{q}{j} \left[\max\left(1 - \frac{jk}{q}, 0 \right) \right]^{q-1} \tag{5.20}$$

where $q = N/2$ if N is even, $q = (N-1)/2$ if N is odd, and κ is the observed value of kappa. Note that the H_0 is rejected if this probability is less than the significance level α. A *post-hoc* review of our data reveals the rejection of H_0 and thus periodicity is present in the time series.

A more recent approach in harmonic analysis of environmental time series is the use of wavelet transforms. A *wavelet* is defined as a family of mathematical functions derived from a basic function (wavelet basis function) by dilation and translation (Chau et al. 2004). As with our discussion on FT, the theory behind wavelet analysis is brief, with detailed summaries provided in more complete sources (Gilliam et al. 2000, Chau et al. 2004, Cooper and Cowan 2008). In general terms, the continuous wavelet transform (CWT) can be defined mathematically as (Kumar and Foufoula-Georgiou 1997)

$$W_f(\lambda, t) = \int_{-\infty}^{\infty} f(u) \bar{\Psi}_{\lambda, t}(u)\, du \qquad \lambda > 0 \tag{5.21}$$

where

$$\Psi_{\lambda, t}(u) \equiv \frac{1}{\sqrt{\lambda}} \Psi\left(\frac{u - t}{\lambda} \right) \tag{5.22}$$

represents the wavelets (family of functions), λ is a scale parameter, t is a location parameter and $\bar{\Psi}_{\lambda,t(u)}$ is the complex conjugate of $\bar{\Psi}_{\lambda,t(u)}$. The inverse can be defined as

$$f(t) = \frac{1}{C_\Psi} \int_{-\infty}^{\infty} \int_{0}^{\infty} \lambda^{-2} Wf(\lambda, u) \psi_{\lambda,u}(t) \, d\lambda \, du \qquad (5.23)$$

where C_Ψ is a constant that depends on the choice of wavelet. Wavelet transforms implemented on discrete values of scale and location are termed discrete wavelet transforms (DWT). Wavelet transforms have advantages over traditional FT for representing functions that have discontinuities, and for accurately deconstructing and reconstructing finite, nonperiodic and/or nonstationary signals of interest (Li and Xie 1997). Their general purpose is in signal decomposition, with the subsequent ability to obtain maximum information stored in the time and space behaviour of the system components.

Consider the study by Labat (2007), which utilizes Morlet wavelet analysis for analyzing annual discharge records of 55 of the world's largest rivers. The Morlet wavelet of arbitrary width and amplitude shown in Figure 5.11 is defined as (Kumar and Foufoula-Georgiou 1997)

$$\Psi(t) = \pi^{-1/4} e^{-i\omega_0 t} e^{-t^2/2} \qquad \omega_0 \geq 5 \qquad (5.24)$$

where Ψ is the wavelet value at nondimensional time t, and ω_0 is the wavenumber.

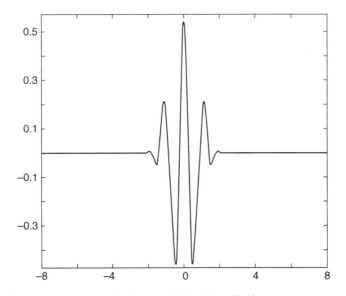

FIGURE 5.11 Morlet wavelet of arbitrary width and amplitude.

The Morlet wavelet tool allowed the investigators to identify statistically meaningful bands of intermittent discharge fluctuations from interannual 5–8-year to decadal, 12–15-year, bidecadal, 28-year fluctuations and 4–70-year. The results proved the usefulness of wavelet-based tools in estimating the temporal evolution of the given hydrological signal's complexity.

A paper by Cho and Chon (2006) further evaluates the application of wavelet analysis in ecological data. They confirm the use of such tools in time-frequency localization and touch upon other applications including the determination of local inhomogeneity in data, and forecasting in time series and in neural network processes for noise removal and smoothing.

5.4   ARIMA MODELS

ARIMA models are a general class of forecasting techniques useful in both stationary and nonstationary time series. Such models have obvious relationships with some of the exponential smoothing techniques in Section 5.3. These similarities are covered in detail throughout this section. In order to begin to understand the ARIMA approach, however, one must first breakdown the acronym into its functional components. These components, both non-seasonal and seasonal, are outlined below.

5.4.1   NONSEASONAL ARIMA MODELS

A nonseasonal ARIMA model is commonly denoted as ARIMA(p,d,q), where p represents the number of autoregressive terms, d is the number of nonseasonal differences, and q is the number of nonseasonal moving average terms.

Note that the autoregressive terms are lags of the *differenced* series appearing in the forecasting equation. Both the ARIMA and seasonal ARIMA (and some of the smoothing models) accommodate a differencing operation that helps to make the time series stationary (exhibiting a fixed mean). A time series that is differenced is said to be an integrated version of the stationary situation. Finally, the lags of the forecast errors are designated moving average terms.

There are a variety of ARIMA models for time series analysis and identifying the one to utilize is a learning process. Identifying the order(s) of differencing needed to produce a stationary time series is the first critical step. Here, the goal is to use the lowest order of differencing that produces a times series that hovers around a well-defined mean, and whose autocorrelation function plot decays fairly rapidly to zero (Hipel 1985). Once the time series has been made stationary, the order(s) of p and q must be identified. Note that each ARIMA(p,d,q) model has corresponding correlation functions. Differencing sometimes has the tendency to introduce negative correlations (pushing the lag -1 autocorrelation to negative values) to a series that had initial positive correlation. Also beware of *overfitting*. The series will not need a higher order of differencing if the lag -1 autocorrelation is zero or negative. With any more than this (e.g., -0.5 or more negative), the series is likely to be over-differenced (Hipel 1985). Also, make note of the standard deviation and relative statistics of the model after differencing as the optimal order is likely the one at which the standard deviation is lowest.

Consider the time series plot (Figure 5.12a) of mean monthly river nitrate levels (mg/l-N) over a 24-month period. This was specified as an ARIMA(0,0,0) model with no differencing, autoregressive, or moving average terms. The time series plot of the residuals is shown in Figure 5.12b, which is simply a plot of deviations from

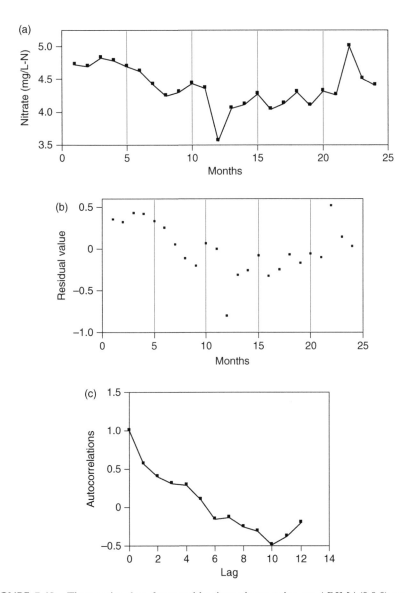

FIGURE 5.12 Time series data for monthly river nitrate using an ARIMA(0,0,0) model: (a) original time series plot, (b) residential plot, and (c) corresponding autocorrelation plot.

the mean. The corresponding autocorrelation plot is shown in Figure 5.12c. Notice the very slow, linear decay pattern, which is standard in a nonstationary time series (Anderson 1979). Model summary statistics are provided in Table 5.10.

It should be clear that some degree of higher-order differencing is needed to allow for a stationary time series. Consider fitting an ARIMA(0,1,0) model to this time series. The ARIMA(0,1,0) is termed a random walk model and contains only a nonseasonal difference and a constant term. The corresponding time series residual plot is shown in Figure 5.13a. Note that the series appears approximately stationary (compared with Figure 5.12b), with no long-term trend. Its corresponding autocorrelation plot shows a slight amount of positive autocorrelation (Figure 5.13b). Summary statistics for the model are listed in Table 5.11 with slight improvements provided by the extra degree of differencing.

Now consider running an ARIMA(1,1,1) model on the time series. The ARIMA(1,1,1) is often referred to as a mixed model where both autoregressive terms and lagged forecast errors have been added. The corresponding time series residual plot is shown in Figure 5.14a. Note that the use of the mixed model approach improved (compared with Figures 5.12b and 5.13a), the stationary nature of the plot with no long-term trends noticeable. The corresponding autocorrelation plot also shows a slight amount of positive correlation (Figure 5.14b). Summary statistics for the mixed model are listed in Table 5.12 with slight improvements provided by the addition of both autoregressive terms and lagged forecast errors. Considering the statistical improvements made by the mixed model and the stationary nature of the plot after ARIMA analysis, forecasting nitrate levels has proved fruitful. This is evident by the generation of a forecasting plot showing a twelve-month nitrate level prediction (Figure 5.15).

Note that there are other nonseasonal ARIMA models that investigators can utilize. An ARIMA(1,1,0), for example, is a differentiated first-order model that adds one lag of the dependent variable to the prediction equation. This is useful if the errors of the random walk model (0,1,0) are autocorrelated. An ARIMA(0,2,1) or

TABLE 5.10
ARIMA(0,0,0) Summary Statistics for Mean Monthly River Nitrate Levels (mg/l-N)

Model/Parameter	Value
ARIMA(0,0,0)	
Sum of squared errors	2.16
Variance estimate	0.09
Standard deviation	0.31
MAPE	5.52
MAE	0.25
AIC	12.4

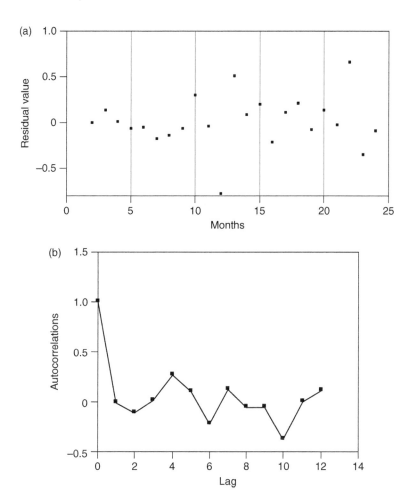

FIGURE 5.13 Time series data for monthly river nitrate concentrations using an ARIMA(0,1,0) model: (a) residual plot and (b) corresponding autocorrelation plot.

(0,2,2) without a constant is termed linear exponential smoothing, which utilizes two nonseasonal differences. The ARIMA(0,2,2) is essentially equivalent to Brown's exponential smoothing model, which predicts that the second difference of the series equals a linear function of the last two forecast errors.

5.4.2 Seasonal ARIMA Models

Section 5.4.1 has examined the use of stationary and nonstationary ARIMA models, and shows how these models can be utilized for forecasting purposes. Although encompassing a wide variety of applications, such models do not adequately address time series that display periodic or seasonal patterns.

TABLE 5.11

ARIMA(0,1,0) Summary Statistics for Mean Monthly River Nitrate Levels (mg/l-N)

Model/Parameter	Value
ARIMA(0,1,0)	
Sum of squared errors	1.76
Variance estimate	0.08
Standard deviation	0.27
MAPE	4.57
MAE	0.18
AIC	8.22

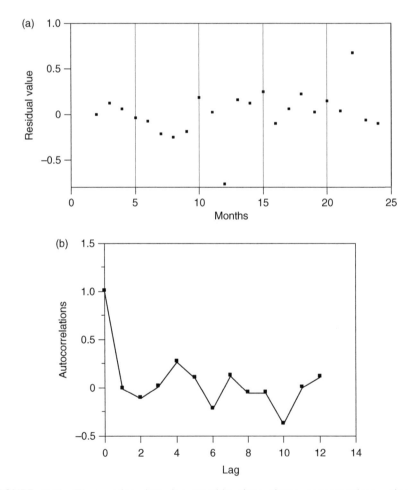

FIGURE 5.14 Time series data for monthly river nitrate concentrations using an ARIMA(1,1,1) model: (a) residual plot and (b) corresponding autocorrelation plot.

TABLE 5.12
ARIMA(1,1,1) Summary Statistics for Mean
Monthly River Nitrate Levels (mg/l-N)

Model/Parameter	Value
ARIMA(1,1,1)	
Sum of squared errors	1.44
Variance estimate	0.07
Standard deviation	0.27
MAPE	3.99
MAE	0.17
AIC	7.86

Seasonal ARIMA models are extensions of the nonseasonal varieties and are commonly denoted as ARIMA(p,d,q)(P,D,Q)s, where P represents the number of seasonal autoregressive terms, D is the number of seasonal differences, Q is the number of seasonal moving average terms, and s represents the period (seasonality) in the data.

The terms p, d, and q are defined in Section 5.4.1. In seasonal ARIMA modeling, the differencing, autoregressive, and moving average components are products of seasonal and nonseasonal polynomials. Reconsider the monthly river flow dataset presented in Table 5.4. Seasonal ARIMA models were appropriately applied to the data. Two particular models are discussed here. Although not conclusive, models were chosen to show distinct differences in summary statistic results. First consider an ARIMA(1,0,1)(1,1,1)12 model with its corresponding forecast plot highlighted in Figure 5.16a. The corresponding residual series plot is shown in Figure 5.16b.

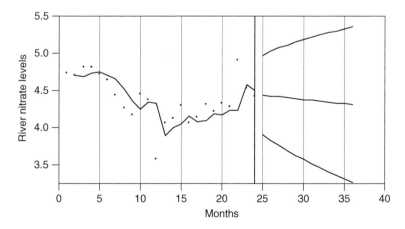

FIGURE 5.15   Forecasting plot of nitrate levels using an ARIMA(1,0,1)(1,1,1)12 model.

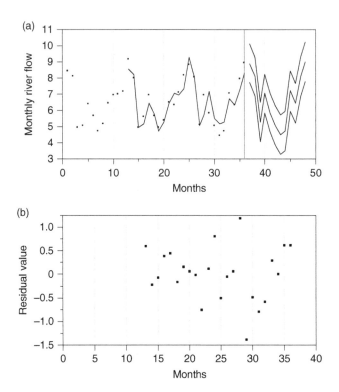

FIGURE 5.16 Monthly river flow time series data employing an ARIMA(1,0,1)(1,1,1)12 model: (a) forecasting plot and (b) corresponding residual plot.

The resultant summary statistics for the ARIMA(1,0,1)(1,1,1)12 are presented in Table 5.13.

Now consider the use of an ARIMA(1,2,1)(1,0,1)12 as a comparison. Note that two additional nonseasonal components were added and a seasonal difference removed. So, should there be obvious effects and would they appropriately model the

TABLE 5.13
ARIMA(1,0,1)(1,1,1)12 Summary Statistics for Mean Monthly River Flow

Model/Parameter	Value
ARIMA(1,0,1)(1,1,1)12	
Sum of squared errors	7.05
Variance estimate	0.37
Standard deviation	0.61
MAPE	6.84
MAE	0.43
AIC	51.0

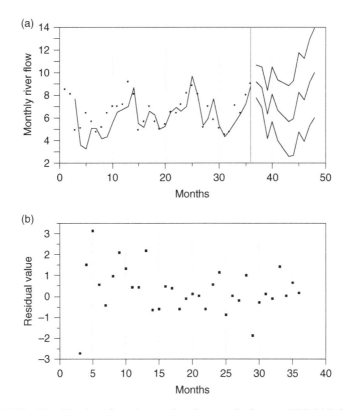

FIGURE 5.17 Monthly river flow time series data employing an ARIMA(1,2,1)(1,0,1)12 model: (a) forecasting plot and (b) corresponding residual plot.

seasonal time series? To answer this, we will need to consider both graphical and tabular examination. Figures 5.17a and b present the forecast plot and corresponding residual series plot, respectively. Pay careful attention to residual plot comparisons. The ARIMA(1,0,1)(1,1,1)12 model produces a residual plot (Figure 5.16b) with values less spread out and hovering closely around zero. This is in direct comparison with the residual plot (Figure 5.17b) for ARIMA(1,2,1)(1,0,1)12. Also consider Table 5.14, which highlights the summary statistics for the ARIMA(1,2,1)(1,0,1)12 model. As discussed previously, both MAE and AIC are key indicators of model performance. Note that these indicators are roughly doubled for the ARIMA(1,2,1)(1,0,1)12 in direct comparison to the ARIMA(1,0,1)(1,1,1)12 model.

5.5 OUTLIERS IN TIME SERIES ANALYSIS

As with any environmental dataset, outliers can arise when performing time series analysis, ultimately leading to biased parameter estimation, poor forecasting, and incomplete or inappropriate decomposition of the series to be studied (Fox 1972). Such outliers are those observations that cannot be properly explained by, for example,

TABLE 5.14

ARIMA(1,2,1)(1,0,1)12 Summary Statistics for Mean Monthly River Flow

Model/Parameter	Value
ARIMA(1,2,1)(1,0,1)12	
Sum of squared errors	16.4
Variance estimate	0.57
Standard deviation	0.75
MAPE	13.08
MAE	0.81
AIC	101.4

ARIMA models. After the initial findings of Fox in 1972, four types of outliers have been proposed (Chen and Liu 1993, Farley and Murphy 1997):

1. Additive outlier (AO)—an observation or pulse that affects (larger or smaller than expected) the time series at a single period only;
2. Innovative outlier (IO)—affects several observations in a time series, often through a dynamic-like pattern;
3. Level shift (LS)—an event that permanently affects the subsequent level of a given time series;
4. Temporary change (TC)—an event that requires a few periods for final decay.

So, why are investigators concerned about outliers in time series and what are the consequences? As discussed in Section 5.3.1, time series model identification is partially based on the estimation of autocorrelations and partial autocorrelations. Therefore, it is expected that outliers likely affect the autocorrelation structure of the given time series. Seminal work by Chen and Liu (1993), for example, show that AOs, TCs, and LSs cause substantial biases in estimated model parameters. Note that IOs show only minor defects. Deutsch et al. (1990) found that the presence of the single outlier results in a true autoregressive model being falsely identified as a moving average model. In addition, identified lag links (p and q) were also found to be skewed.

EXAMPLE PROBLEM 5.2

1. Briefly discuss how the different outliers can affect the forecasts of ARIMA models.

Answer: Studies have shown that AOs can cause misleading prediction intervals and inflate the estimated variance of the given time series (Holta 1993). Level shifts and TCs may also increase the width of prediction intervals and have serious effects on a point forecast even when outliers are not close to the forecast origin (Trívez 1993).

There are numerous methods used for time series outlier (single and multiple) detection, with full details beyond the scope of this book. However, methods such as likelihood ratio tests, tests based on rank transformations, estimates of autocorrelation and partial autocorrelation functions are often employed. In addition, studies have shown the usefulness of wavelet-based analysis for outlier detection (Greenblatt 1994).

5.6 EXPORT COEFFICIENT MODELING

Numerous models are available to assess nonpoint pollution and sediment sources, their delivery to surface waters, and to aid the forecasting of (in collaboration with time series analysis), future events in terms of pollution loading. The *export coefficient modeling* approach, for example, predicts nutrient loading (annual and seasonal) into *watersheds*. This approach considers the complexity of land-used systems and the spatial distribution of land-use data from various nutrient sources and predicts total loads into a particular watershed (Johnes 1996).

The export coefficient model for predicting nutrient loading into receiving waters was initially developed to determine the origin of increased nutrients in North American lakes (Omernik 1976, Beaulac and Reckhow 1982). This model was later described in detail as the agricultural watershed scale (Gburek et al. 1996, Johnes 1996, Johnes et al. 1996). It takes into account the complexity of land-use systems and allows scaling-up from plot scale to a larger watershed size. *Export coefficients* express the rate of nutrient loss from each identifiable source. In terms of land-use, export coefficients express the rate at which nutrients are exported from each land type present in a given watershed. Examples of phosphorus export coefficients ($kg\ ha^{-1}\ yr^{-1}$) for land types are presented in Table 5.15 (Johnes 1996). Key factors regarding the selection of land-use export coefficient values include basin geology, hydrological pathway, and land management practices. Exports will also be influenced by the use of organic animal waste and inorganic fertilizers.

Table 5.16 lists example export coefficients for the different animal types and sewage (human) population equivalents for both sewage treatment works (STWs) and private septic systems. For animals, export coefficients are derived from the proportion of phosphorus loss from wastes voided from stock houses and grazing land. Note that the model assumes standard losses, for example, 3.00% for sheep, derived from Vollenweider (1968) as the mean value of best-case (1.00%) and worst-case (5.00%) scenarios. Human export coefficients reflect the use of phosphate-rich detergents and dietary factors, and are adjusted to take into account primary and secondary sewage treatment or the use of septic tanks (Johnes 1996).

The term *watershed* refers to a geographic area of land that drains downhill to common points including rivers, streams, ponds, lakes, wetlands, and estuaries.

The export coefficient model has been successfully used as a management tool by accurate prediction of nutrient loading, and then by allowing informed decisions

TABLE 5.15

Example Phosphorus Export Coefficients for Landcover Types

Landcover Type	Export Coefficient (kg ha^{-1} yr^{-1})
Sea/estuary	0.00
Inland water	0.00
Beach and coastal bare	0.00
Salt marsh	0.00
Grass heath	0.02
Mown/grazed turf	0.20
Meadow/verge/semi-natural	0.20
Rough/marsh grass	0.02
Bracken	0.02
Dense shrub heath	0.02
Scrub/orchard	0.02
Deciduous woodland	0.02
Coniferous woodland	0.02
Tilled land	0.66
Suburban/rural development	0.83
Continuous urban	0.83
Inland bare ground	0.70
Lowland bog	0.00
Open shrub heath	0.02
Unclassified	0.48

Source: Johnes, P. J., *Journal of Hydrology*, 183, 323–349, 1996.

TABLE 5.16

Example of Phosphorus Export Coefficients for Animal Types and Sewage Population Equivalents

Phosphorus Source	Export Coefficients
Animals	
Cattle	6.9 g head^{-1} yr^{-1}
Horses	4.4 g head^{-1} yr^{-1}
Pigs	20.0 g head^{-1} yr^{-1}
Sheep	4.4 g head^{-1} yr^{-1}
Humans	
STWs	0.38 kg ca^{-1} yr^{-1}
Septic systems	0.24 kg ca^{-1} yr^{-1}

to be taken on which measures will be most effective in reducing loads into receiving waterbodies. In a management context, this model can be used to predict nutrient loading on a watershed scale when certain parameters (i.e., coefficients) are varied. As expected, though, this model is not without its flaws and limitations. Chapter 1 has discussed the nature of phosphorus dynamics, which ultimately leads to difficulty in monitoring and modeling. Many strategies may be narrowly targeted, for example, due to phosphorus transport characteristics. This can finally result in inconsistent management decisions (Gburek et al. 2000), especially at the watershed scale.

Uncertainties also exist when considering spatial heterogeneity within given land-use classes. Even watersheds with similar land-use patterns have been shown to exhibit export coefficients that are highly variable (Khadam and Kalvarachchi 2006). Hydrologic variability also plays an important role in export coefficient uncertainty. This is discussed in detail by Sharpley et al. (2002) who show that phosphorus loads from a watershed are directly related to the volume of surface runoff. Khadam and Kalvarachchi (2006) developed an erosion-scaled export coefficient approach to account accurately for the variability of annual phosphorus loading due to hydrologic variability. In this unique approach, sediment discharge from runoff processes is introduced into the model as a surrogate for hydrologic variability.

5.7 EXPANDED RESEARCH APPLICATION IV—EXPORT COEFFICIENT MODELING AND TIME SERIES ANALYSIS OF PHOSPHORUS IN A CHALK STREAM WATERSHED

Phosphorus, transported via a wide range of point and nonpoint sources, is known to affect the overall quality of river systems. Land-use, particularly agriculture, is considered to be a significant nonpoint source of phosphorus export from watersheds (Dillon and Kirchner 1975). The major point sources are sewage and industrial discharges, which also contribute to overall phosphorus export (Muscutt and Withers 1996). Cumulatively, phosphorus export (loading to the watershed) can result in eutrophication, sometimes leading to an increase in biomass and primary productivity within aquatic communities (Young et al. 1999).

5.7.1 STATEMENT OF PROBLEM

An appraisal of water quality is needed before management practices are implemented that minimize the likelihood of eutrophication. Assessment has traditionally involved a physical approach that considers real-time changes based on contemporary measurements and small-scale plot size (Moss et al. 1996). The export coefficient model has proved to be a simple but effective model in predicting phosphorus loading from point and nonpoint sources. Consider the following study by Hanrahan et al. (2001a).

5.7.2 RESEARCH OBJECTIVES

The investigators examined the hypothesis that the export coefficient modeling approach could be used to predict TP loading in the River Frome watershed (Dorset, UK) on both an annual and seasonal (monthly) basis from 1990–1998. Predicted values were compared with historical TP and river flow time series data.

5.7.3 EXPERIMENTAL METHODS AND CALCULATIONS

The River Frome starts on the North Dorset Downs near Evershot and flows into Poole Harbor in the south-west of the UK (Figure 5.18). The most important geological formation is chalk, which comprises nearly 50% of the 41,429 ha watershed (Casey and Newton 1973). The watershed has a braided network of channels, both naturally occurring and constructed for flood relief. The soils are shallow and well drained with a few areas of clay-influenced soils in the lower watershed. Land-use in the watershed is predominately meadow/verge/semi-natural (31.0%), tilled land (28.0%) and mown/grazed turf (19.4%).

The general equation for annual phosphorus loading used in this study (Johnes 1996) was

$$PL = \sum_{i=1}^{n} (A_i \times E_i) \qquad (5.25)$$

where PL is the TP load into the watershed; A is the area (hectares) of land type within the watershed i, or number of animal types i, or sewage population equivalents (PE) i; E_i is the annual export coefficient for each source i (kg ha^{-1} yr^{-1} (land types), kg ca^{-1} yr^{-1} (population equivalents) and g head^{-1} yr^{-1} (animal)). For land-use and humans, coefficients were simply multiplied by each particular land type and sewage population equivalents (STWs and private septic systems). For animals, the TP load was determined from the quantity of waste produced (kg) per kg of live animal

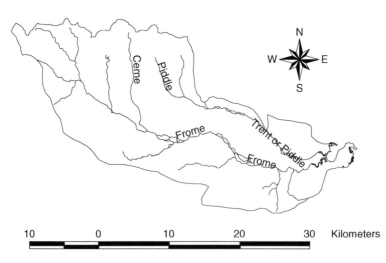

FIGURE 5.18   River Frome watershed map, Dorset, UK. (From Hanrahan, G., et al., *Journal of Environmental Quality*, 30, 1738–1746, 2001. With permission.)

weight (Vollenweider 1968), the total livestock population, and the animal export coefficients for phosphorus.

Seasonal TP loading from land-use in the watershed was estimated using the equation from May et al. (2001):

$$E_j = B_j c_b + \frac{Q_j - B_j}{Q_T - B_T} \sum_i E_i A_i \qquad (5.26)$$

where E_j is the TP load per month (j), i denotes the landcover type, A_i is the area of each landcover type, E_i is the annual export for the ith landcover type, Q_j is the discharge of water from the watershed during month j, B_j is the discharge of water from baseflow during the month j, Q_T is the annual discharge of water, and B_T is the annual contribution of baseflow with a fixed phosphorus concentration of c_b. Note that for both annual and seasonal studies, the export coefficients listed in Tables 5.15 and 5.16 were utilized. Historical TP and filterable reactive phosphorus (FRP) concentrations were determined weekly from the River Frome (East Stoke location) and subsequently used in the modeling process.

5.7.4 RESULTS AND INTERPRETATION

The annual observed TP load for 1998 (23,400 kg yr^{-1}) at East Stoke was calculated by multiplying the mean daily discharge for a given sampling date by the TP concentration measured on that date. Due to the variation in time intervals between measurements (3–5 times monthly), the observed annual load was obtained using the following equation (Clasper 1999):

$$l = \frac{\sum \Delta t l}{\sum \Delta t} \times 365 \qquad (5.27)$$

where $\sum \Delta t l$ is the sum of loads calculated for each time interval, divided by $\sum \Delta t$ (sum of time intervals) and multiplied by 365 (days).

The modeled annual TP loading from both point and nonpoint sources for 1998 was calculated to be 25,605 kg yr^{-1}. Land use was the primary factor, with nonpoint sources contributing the greatest amount at 14,085 kg yr^{-1}. This is not surprising as the cultivation of crops and the subsequent addition of fertilizers have increased markedly in the UK from 1970 onward (Johnes 1996). In the United States, between 1945 and 1993, the use of phosphorus-containing fertilizers increased from 0.5 million to nearly 1.8 million metric tons per year (Kornecki et al. 1999). Table 5.17 lists the landcover types in terms of % area of the watershed and individual contributions of TP (kg yr^{-1}). As shown, tilled land (7666 kg yr^{-1}), meadow/verge/semi-natural (2562 kg yr^{-1}), mown/grazed turf (1611 kg yr^{-1}) and suburban/rural development (1486 kg yr^{-1}) accounted for the majority of the delivered TP to the River Frome watershed. Other studies have shown that watersheds with intensive agriculture (tilled land) export high levels of phosphorus. Ahl (1972) reported a TP export of 9200 kg yr^{-1} from an intensively farmed watershed, nearly two times higher than watersheds dominated by forestland. Gächter and Furrer (1972) compared phosphorus export from agricultural watersheds in the Lower Alps and Swiss lowlands with

TABLE 5.17

Land Cover Types and Their Contribution of Phosphorus in the River Frome Watershed

Land Cover Type	% of Watershed Area	Phosphorus Export (kg P yr^{-1})
Sea/estuary	0.00	0
Inland water	0.02	0
Beach and coastal bare	0.00	0
Salt marsh	0.00	0
Grass heath	1.27	10.5
Mown/grazed turf	19.4	1611
Meadow/verge/semi-natural	31.0	2563
Rough/marsh grass	0.98	8.10
Bracken	0.0005	0.004
Dense shrub heath	0.50	4.22
Scrub/orchard	0.92	7.64
Deciduous woodland	7.37	61
Coniferous woodland	2.64	21.9
Tilled land	28.0	7666
Suburban/rural development	4.32	1486
Continuous urban	0.19	65.8
Inland bare ground	0.52	151
Lowland bog	0.10	0
Open shrub heath	0.63	5.22
Unclassified	2.13	424
Total	100	14,085

Source: Hanrahan, G., et al., *Journal of Environmental Quality*, 30, 1738–1746, 2001. With permission.

forested watersheds in the same region. Results showed that there was at least a 10-fold increase of phosphorus export in agricultural-dominated watersheds. A study of thirty-four watersheds in Southern Ontario (Dillon and Kirchner 1975) found, in general, higher phosphorus exports from forested watersheds containing some agricultural activities, compared with those that were predominately forested. In addition, studies from the United States (Sonzogni and Lee 1972, Sonzogni et al. 1980) have reported high levels of phosphorus export from watersheds with a high intensity of tilled land.

Table 5.18 lists the results of the modeled animal and human exports. There were eight STWs: Dorchester, Wool, Bradford Peverell, Warmwell, Maiden Newton, Broadmayne, Tollerporcorum, and Evershot located within the watershed boundaries, accounting for a total of 8869 kg P yr^{-1}. There were an additional 5548 PE on private septic systems throughout, accounting for an annual contribution of 10,200 kg yr^{-1} (8869 kg yr^{-1}, STWs; 1331 kg yr^{-1}, septic systems). The remaining 1320 kg yr^{-1} arose from the four animal types (pigs = 713, cattle = 330, sheep = 250, horses = 28).

TABLE 5.18
Phosphorus Export from STWs and Animals in the River Frome Watershed

STWs/Animal Types	Population Equivalents (P.E.)/ Number of Animals	Export (kg P yr^{-1})
Dorchester	13,007	4943
Wool	5001	1900
Bradford Peverell	1600	608
Warmwell	1410	536
Maiden Newton	1005	382
Broadmayne	860	326
Tollerporcorum	250	95
Evershot	210	79
Total	23,340	8869
Pigs	35,585	713
Cattle	48,000	330
Sheep	56,599	250
Horses	640	28
Total	140,824	1320

Source: Hanrahan, G., et al., *Journal of Environmental Quality*, 30, 1738–1746, 2001. With permission.

Annual exports of TP are not always the best predictors of whether an ecosystem will be highly productive in terms of biomass (Reckhow and Simpson 1980, Bock et al. 1999). Thus, assessing seasonal trends of TP and the relationship between TP and FRP within the watershed are important because large proportions of TP are delivered during periods of high discharge (Haygarth et al. 1998), and FRP is the fraction of TP most readily available for uptake by aquatic plants. In this study, the highest TP and FRP concentrations occurred at the highest river discharge, in the autumn of 1998 as shown by the time series plot (Figure 5.19). Regression analysis of TP concentration versus discharge (Figure 5.20) from 1990–1998 illustrates this pattern, with a significant relationship ($r^2 = 0.68$) and the highest concentrations of TP occurring at the highest discharge. In another extended study covering eight years, Ulén (1998) found that the twelve wettest months of this time period (representing 12.5% of total time) contributed more than a third of the TP export to various Swedish watersheds. Similar trends have been observed in other rivers and streams: Duffin Creek, Canada (Hill 1981); Lakeland streams, UK (Lawlor et al. 1998); Dartmoor streams, UK (Rigler 1979); Karkloof, Magalies, Vaal and Vet Rivers, South Africa (Meyer and Harris 1991); Foron River, France (Dorioz et al. 1998); and River Cherwell, UK (May et al. 2001) where TP loads generally increased with increasing river discharge.

An examination of observed monthly TP concentrations for 1998 and for the extended time period of 1990–1997 was undertaken (Figure 5.21). Results for 1998 revealed elevated values above the yearly mean (7.16 µM) during January, March, June, August, September, and December (7.58–8.14 µM). The extended time period of 1990–1997 revealed greater seasonal differences. Values above the mean (7.58 µM)

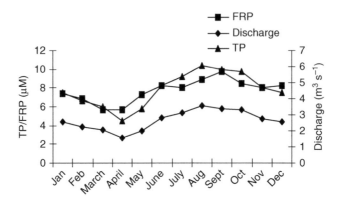

FIGURE 5.19   Seasonal time series plot of TP and FRP concentrations and river flow at East Stoke Station, River Frome, UK (1990–1998). (From Hanrahan, G., et al., *Journal of Environmental Quality*, 30, 1738–1746, 2001. With permission.)

were observed during January, June, and August–December (8.2–9.78 μM), with September recording the highest mean monthly value. April showed the lowest mean monthly value (5.52 μM).

Observed FRP data was also examined for seasonal trends for the period 1990–1997. Results showed a similar trend as that for TP, with the highest mean monthly values recorded in September and October (5.68 μM). The lowest mean monthly value occurred in April (2.87 μM). Seasonal trends for 1998 could not be assessed due to unavailability of data. Regression of FRP/TP ratios with corresponding river discharges (Figure 5.22) showed a correlation of 0.43. The highest mean FRP/TP ratio (0.71) corresponded with the highest mean discharge (8.40 m³ s⁻¹) in the month of January. April showed the lowest FRP/TP ratio (0.52) and a correspondingly low

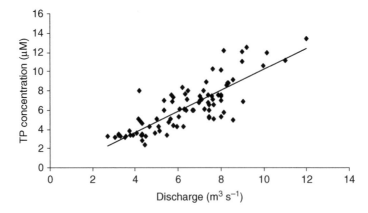

FIGURE 5.20   TP versus flow in the River Frome from 1990–1998. (From Hanrahan, G., *Catchment scale monitoring and modeling of phosphorus using flow injection analysis and on export coefficient model*, Ph.D. thesis, University of Plymouth, UK, 2001b.)

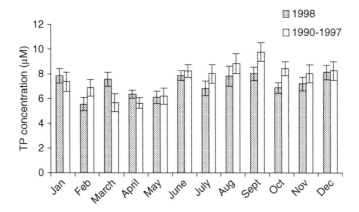

FIGURE 5.21 Observed monthly TP concentrations in the River Frome (1990–1998). Error bars = 3*s*. (From Hanrahan, G., *Catchment scale monitoring and modeling of phosphorus using flow injection analysis and on export coefficient model*, Ph.D. thesis, University of Plymouth, UK, 2001b.)

discharge of 4.24 m^3 s^{-1}. This suggests that FRP is not only controlled by the dilution of point inputs but that nonpoint inputs are also particularly important during storm events. Similar trends in the FRP/TP ratio were seen in the River Cherwell watershed in Oxfordshire, UK with values ranging from 0.50 to 0.93 (May et al. 2001). The latter value, slightly higher than the highest mean ratio (0.71) for the River Frome, was taken just below a major STW.

In order to predict monthly TP loading, it was first necessary to determine baseflow water discharges, as they are an integral part of Equation 5.26. This was done by averaging discharges for days subsequent to a period of at least five days of no rainfall. Baseflow discharges also showed a seasonal dependence with the highest

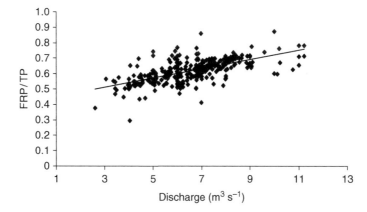

FIGURE 5.22 FRP/TP ratio versus river flow in the River Frome (1990–1998). (From Hanrahan, G., *Catchment scale monitoring and modeling of phosphorus using flow injection analysis and on export coefficient model*, Ph.D. thesis, University of Plymouth, UK, 2001b.)

mean monthly value recorded during January ($7.83 \, m^3 \, s^{-1}$). The lowest mean monthly value ($1.08 \, m^3 \, s^{-1}$) occurred in September.

Results for the prediction of monthly TP loading from Equation (5.26) are shown in Figure 5.23. In these calculations, the export coefficients from Table 5.16 were used, as these provided the best annual estimate, along with annual TP inputs from both animal and human sources. Animal sources were also calculated using Equation 5.25, whereas human sources were divided equally among the 12 months ($833 \, kg \, month^{-1}$). The predicted results show a seasonal dependence similar to that for the observed 1998 data. December showed the highest monthly TP loading at 2367 kg (including human/animal sources) with April recording the lowest loading at 1,778 kg. Large differences between observed and predicted phosphorus loadings were found for the months of February and May with predicted values 37% and 24%, respectively, greater than the observed values. For September, the predicted value was 14% lower than the observed value. The annual level of rainfall was 1003 mm, with a mean monthly level of 84 mm across the watershed. However, February and May recorded levels were above the mean at 119 mm and 153 mm, respectively.

In general, FRP and TP followed a similar seasonal trend but the FRP/TP ratio was not constant. Highest ratios were observed in the winter months, when river discharges were higher, while the lowest ratios occurred in the summer months. Internal processes play a major role in the uptake and release of phosphorus. During spring/summer, TP decrease can be explained by uptake by bottom sediments and biological uptake, and, in the case of the River Frome, which is a chalk stream, there is a possibility of coprecipitation of phosphate with calcite (House et al. 1986, Jansson 1988). The opposite can occur in autumn/winter months, when release is attributed to dying plants and algae, and resuspension of bottom sediments from increasing

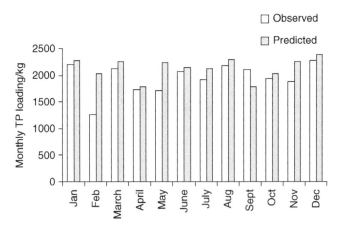

FIGURE 5.23　Predicted versus observed seasonal TP loading in the River Frome watershed (1998). (From Hanrahan, G., *Catchment scale monitoring and modeling of phosphorus using flow injection analysis and on export coefficient model*, Ph.D. thesis, University of Plymouth, UK, 2001b.)

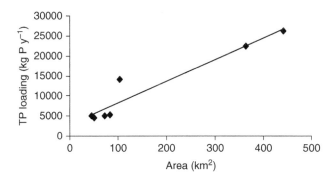

FIGURE 5.24   Relationship between watershed area (km²) and TP loading (kg yr⁻¹) in terms of land use. (From Hanrahan, G., *Catchment scale monitoring and modeling of phosphorus using flow injection analysis and on export coefficient model*, Ph.D. thesis, University of Plymouth, UK, 2001b.)

river discharge (Dorioz et al. 1989, House et al. 1998). Knowledge of trends in TP loading could therefore provide useful information on the likely variability in FRP concentrations in the Frome watershed, at least on a monthly time scale.

Consider the comparison with other watersheds. Regression analysis of TP loading from land-use (kg yr⁻¹) on the watershed area (km²) for the Foron River, France (Dorioz et al. 1998), (83 km², 5400 kg yr⁻¹), Frome (414 km², 14,085 kg yr⁻¹) and five other UK river watersheds (Johnes and Hodgkinson 1998): Slapton Ley (46 km², 1970 kg yr⁻¹), Ryburn (51 km², 4540 kg yr⁻¹), Esk (72 km², 5083 kg yr⁻¹), Waver (104 km², 14,040 kg yr⁻¹) and Windrush (363 km², 6500 kg yr⁻¹) showed a linear relationship, $r^2 = 0.926$ (Figure 5.24). This is in agreement with reported data on 108 watersheds with similar land-use (Prairie and Kalff 1986). This linear relationship between watershed export and area strongly suggests that phosphorus export can be estimated solely from land area. However, the majority of the watersheds assessed to date are mixed agricultural watersheds and assessment of model performance in watersheds dominated by urban environments, intensive farming, and upland areas is required before firm conclusions can be drawn with respect to the generic application of the model.

The Frome watershed shows a greater contribution of phosphorus from sewage input (10,200 kg yr⁻¹, 40% of total) compared with the Windrush (3780 kg yr⁻¹, 17% of total) and Slapton (780 kg yr⁻¹, 16% of total), mainly due to population differences. Owens (1970) estimated that 1.5 persons/hectare of watershed could contribute 31% of the TP load, similar to the value obtained for this study (38%) at 1.3 persons/hectare.

The sensitivity of the model was assessed to determine the factors that exert the greatest influence on the prediction of phosphorus export. Each export coefficient was adjusted by an arbitrary value of ±10%, while holding all others at the base coefficient value. The results of the sensitivity analysis are presented in Table 5.19 and expressed as a percentage change from the base contribution. As shown, the analysis suggests that nutrient export from STWs (3.5%), tilled land (2.7%), meadow/verge/

TABLE 5.19
Sensitivity Analysis for the River Frome Watershed

Nutrient Source	% Change from Base Contribution
Humans: STWs	3.5
Tilled land	2.7
Meadow/verge/semi-natural	1.0
Mown/grazed turf	0.62
Suburban/rural development	0.60
Humans: septic systems	0.52
Pigs	0.30
Unclassified land	0.16
Cattle	0.13
Sheep	0.10
Continuous urban	0.03
Deciduous woodland	0.03
Horses	0.01
Coniferous woodland	0.009
Grass heath	0.004
Rough/marsh grass	0.001
Inland bare ground	0.001
Scrub/orchard	0.0009
Open shrub heath	0.0006
Dense shrub heath	0.0005
Bracken	0.00003

Source: Hanrahan, G., et al., *Journal of Environmental Quality*, 30, 1738–1746, 2001. With permission.

semi-natural (1.0%) and mown/grazed turf (0.6%) are the most important single factors controlling the export from the Frome watershed.

Export coefficients for human and animal sources were taken solely from Johnes (1996) and judged to be an appropriate reflection of literature values. Furthermore, human sources were separated into both private septic systems and STWs, thus allowing additional characterization of TP loading. Predicted values were compared with observed (measured) data from 1998. The final calibrated model prediction (25,605 kg yr^{-1}) was within 9% of the observed TP load (23,400 kg yr^{-1}).

5.7.5 SUMMARY AND SIGNIFICANCE

The export coefficient model allowed the prediction of TP loading, both monthly and annually, in the Frome watershed, UK. A total annual TP load of 25,605 kg yr^{-1} was calculated using the model, which compared to an observed (measured) value of 23,400 kg yr^{-1}. Nonpoint sources (land use, animals and septic tanks) made the most significant contribution to the TP load (65%), with 35% of TP coming from STWs. Predicted TP loadings were consistent with other published data in terms of loading per watershed area. An analysis of the FRP to TP ratio shows that the ratio varied with discharge, and that TP could therefore be used to obtain valuable information on FRP concentrations in the watershed. The results of the sensitivity analysis show that STWs (3.5%), tilled land (2.7%), meadow/verge/semi-natural (1.0%) and mown and grazed turf (0.6%) had the most significant effect (percent difference from base contribution) on model prediction.

Seasonal results showed that, on the whole, the predicted monthly TP loads provided a good estimate of actual TP loads. However, important differences were observed in months of variable discharge (i.e., February and May). The differences appear, in part, to be due to an underestimation of the observed TP loading resulting from the sampling program missing some storm events. In addition, it is likely that TP is retained in the river during spring and summer due to uptake by bottom sediments and plant and algal growth, thus helping to account for lower observed values. September was the only month when observed export was greater than predicted. This could be explained by the generation of in-stream TP remobilization from sediment porewaters, resuspension of sediments, and degradation of algae and plants in the river. Overall, the observed database was limited (3–5 samplings per month) and sampling dates in these months were on days of lower river discharges. For example, the mean discharge for the dates sampled during February and May 1998 was $3.94 \, m^3 \, s^{-1}$. However, mean discharge for all days during the same months was greater ($5.11 \, m^3 \, s^{-1}$). These discrepancies were most likely due to unrepresentative sampling, and illustrate the potential use for a model of this type to design and coordinate representative sampling campaigns.

5.8   CHAPTER SUMMARY

The importance of time series analysis and related smoothing and digital filtering techniques has been thoroughly examined with the use of environmentally based applications. Both nonseasonal and seasonal models were explored and detailed information and interpretation presented in graphical and tabular formats. The use of time series models in relation to the export coefficient modeling approach was also considered in detail. This approach was shown to have the ability to predict effectively the nutrient loading into watersheds on both an annual and seasonal basis. In combination, time series analysis and export coefficient modeling can provide effective prediction of pollutant loads in given watershed processes.

5.9   END OF CHAPTER PROBLEMS

1. Describe the benefits of smoothing and digital filtering in time series analysis.
2. Given the following weekly river discharge ($m^3 \, s^{-1}$) data, determine the appropriate smoothing method for this time series. Provide logical reasoning for your choice.

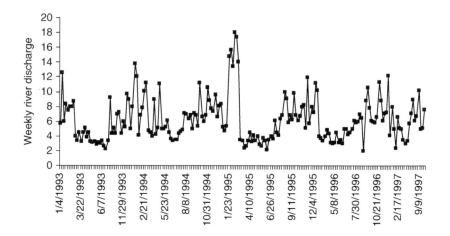

3. In Example Problem 5.2, describe how one may identify the best values for the smoothing parameters chosen to forecast the next few values in the river discharge series.

4. Discuss the major differences between the Fourier analysis method and wavelet-based analysis in terms of decomposition of a time series. What are the major advantages and disadvantages of each?

5. Describe the distinguishing differences between nonseasonal and seasonal time series models.

6. In Section 5.5, discussion on how time series outliers present difficulties when performing a forecasting procedure is presented. Provide logical reasoning why outliers near the beginning of the forecast period can be especially problematic.

7. As discussed in the chapter, most formal studies of the time series models are concerned with stationary concepts. Explain why this stationary nature is important in model development and interpretation.

8. Discuss why high resolution monitoring of nutrients would be beneficial in developing and refining export coefficient models for watershed loading purposes.

REFERENCES

Ahl, T. 1972. Plant nutrients in Swedish lake and river water. *Verhhandlungen Internationaler Vereinigung fuer Theoretische und Angewandte Limnologie* 18: 362–369.

Akaike, H. 1974. A new look at the statistical model identification. *IEEE Transactions on Automatic Control* 19: 716–723.

Anderson, O. D. 1979. On realization from nonstationary time processes. *The Journal of the Operational Research Society* 30: 253–258.

Astel, A., Mazerski, J., Polkowska, Z., and Namieśnik, J. 2004. Application of PCA and time series analysis in studies of precipitation in Tricity (Poland). *Advances in Environmental Research* 8: 337–349.

Athanasopoulos, G. and Hyndman, R. J. 2008. Modelling and forecasting domestic tourism. *Tourism Management* 29: 19–31.

Beaulac, M. N. and Reckhow, K. H. 1982. An examination of land use-nutrient export relationships. *Water Research Bulletin* 18: 1013–1023.

Blair, G. M. 1995. A review of discrete Fourier transform. 1. Manipulating the powers of two. *Electronics and Communications Engineering Journal* 7: 169–177.

Bock, M. T., Miller, B. S., and Boman, M. T. 1999. Assessment of eutrophication in the Firth of Clyde: Analysis of coastal water data from 1982–1996. *Marine Pollution Bulletin* 38: 222–231.

Box, G. and Jenkins, G. M. 1970. *Time Series Analysis: Forecasting and Control.* San Francisco: Holden–Day.

Brown, R. G. 2004. *Smoothing, Forecasting and Prediction in Discrete Time Series.* Mineola, NY: Courier Dover Publications.

Casey, H. and Newton, P. V. R. 1973. The chemical composition and flow of the River Frome and its main tributaries. *Freshwater Biology* 3: 317–333.

Chaloupka, M. 2001. Historical trends, seasonality and spatial synchrony in green sea turtle egg production. *Biological Conservation* 101: 263–279.

Chau, F. T., Liang, Y. Z., Gao, J., and Shao, X. G. 2004. *Chemometrics: From Basics to Wavelet Transform.* Hoboken: John Wiley & Sons, Inc.

Chen, C. and Liu, L. 1993. Joint estimation of model parameters and outlier effects in time series. *Journal of the American Statistical Association* 88: 284–297.

Cho, E. and Chon, T.-S. 2006. Application of wavelet analysis to ecological data. *Ecological Informatics* 1: 299–233.

Clasper, P. 1999. Loading and fluxes of nutrients in the freshwater Tamar and tributaries and freshwater Teign watersheds. BSc (Hons) thesis, University of Plymouth.

Cooley, J. W. and Tukey, O. W. 1965. An algorithm for the machine calculation of complex Fourier series. *Mathematics of Computation* 19: 297–301.

Cooper, G. R. J. and Cowan, D. R. 2008. Comparing time series using wavelet-based semblance analysis. *Computers & Geosciences* 34: 95–102.

Deutsch, S. J., Richards, J. E., and Swain, J. J. 1990. Effects of a single outlier on ARMA identification. *Communications in Statistics, Theory and Methods* 19: 2207–2227.

Dillon, P. J. and Kirchner, W. B. 1975. The effects of geology and land use on the export of phosphorus from watersheds. *Water Research* 9: 135–148.

Dilmaghani S., Henry, I. C., Soonthornnonda, P., Christensen, E. R., and Henry, R. C. 2007. Harmonic analysis of environmental time series with missing data or irregular spacing. *Environmental Science & Technology* 41: 7030–7038.

Dorioz, J. M., Cassell, E. A., Orand, A., and Eisenman, K. G. 1998. Phosphorus storage, transport and export dynamics in the Foron River watershed. *Hydrological Processes* 12: 285–309.

Farley, E. V. and Murphy, J. M. 1997. Time series analysis: Evidence for management and environmental influences on Sockeye Salmon catches in Alaska and Northern British Columbia. *Alaska Fisheries Research Bulletin* 4: 36–53.

Fox, A. J. 1972. Outliers in time series. *Journal of the Royal Statistical Society, Series B,* 34: 350–363.

Gardner, E. S. 1985. Exponential smoothing: the state of the art. *Journal of Forecasting* 4: 1–28.

Gächter, R. and Furrer, O. J. 1972. Der beitrag der landwirtschaft zur etrophierung der Gëwasser in der Schweiz. *Schweizerische Zeitschrift fuer Hydrologie* 34: 41–70.

Gburek, W. J., Sharpley, A. N., and Poinke, H. B. 1996. Identification of critical source areas for phosphorous export from agricultural watersheds. In *Advances in Hillslope Processes*, ed. M. G. Anderson et al. Chichester, UK: John Wiley & Sons.

Gburek, W. J., Sharpley, A. N., Heathwait, L., and Folmar, G. J. 2000. Phosphorous management at the watershed scale: A modification of the phosphorous index. *Journal of Environmental Quality* 29: 130–144.

Gilliam, X., Dunyak, J., Doggett, A., and Smith, D. 2000. Coherent structure detection using wavelet analysis in long time-series. *Journal of Wind Engineering and Industrial Aerodynamics* 88: 183–195.

Greenblatt, S. A. 1994. Wavelets in econometrics: an application to outlier testing. *Working Paper*, University of Reading, UK.

Groves-Kirkby, C. J., Denman, A. R., Crockette, R. G. M., Phillips, P. S., and Gillmore, G. K. 2006. Identification of tidal and climatic influences within demostic radon time-series from Northamptonshire, UK. *Science of the Total Environment* 367: 191–202.

Hanrahan, G., Gledhill, M., House, W. A., and Worsfold, P. J. 2001a. Phosphorus loading in the Frome catchment, UK: seasonal refinement of the coefficient modeling approach. *Journal of Environmental Quality* 30: 1738–1746.

Hanrahan, G. 2001b. *Catchment scale monitoring and modeling of phosphorus using flow injection analysis and on export coefficient model*, Ph.D. thesis, University of Plymouth, UK.

Haygarth, P. M., Hepworth, L., and Jarvis, S. C. 1998. Form of phosphorus transfer and hydrological pathways from soil under grazed grassland. *European Journal of Soil Science* 49: 65–72.

Hill, A. R. 1981. Stream phosphorus exports from watersheds with contrasting land uses in southern Ontario. *Water Resources Bulletin* 17: 627–634.

Hipel, K. W. 1985. Time series analysis in perspective. *Water Resources Bulletin* 21: 609–624.

Hirji, K. F. 1997. A review and a synthesis of the fast Fourier transform algorithms for exact analysis of discrete data. *Computational Statistics and Data Analysis* 25: 321–336.

Hogrefe, C., Vempaty, S., Rao, S., and Porter, P. S. 2003. A comparison of four techniques for separating different time scales in atmospheric variables. *Atmospheric Environment* 37: 313–325.

Holta, L. K. 1993. The effect of additive outliers on the estimates from aggregated and disaggregated ARIMA model. *International Journal of Forecasting* 9: 85–93.

House, W. A., Casey, H., and Smith, S. 1986. Factors affecting the coprecipitation of inorganic phosphate with calcite in hardwaters—II. Recirculating experimental stream system. *Water Research* 20: 923–927.

House, W. A., Jickells, T. D., Edwards, A. C., Praska, K. E., and Denison, F. H. 1998. Reactions of phosphorus with sediments in fresh and marine waters. *Soil Use and Management* 14: 139–146.

Jansson, M. 1988. Phosphate uptake and utilization by bacteria and algae. *Hydrobiologia* 170: 177–189.

Johnes, P. J. 1996. Evaluation and management of the impact of land use change on the nitrogen and phosphorus load delivered to surface waters: the export coefficient modelling approach. *Journal of Hydrology* 183: 323–349.

Johnes, P. J. and Hodgkinson, R. A. 1998. Phosphorus loss from agricultural watersheds: pathways and implications for management. *Soil Use and Management* 14: 175–185.

Johnes, P. J., Moss, B., and Phillips, G. 1996. The determination of total nitrogen and total phosphorous concentrations in freshwaters from land use, stock headage and population data: testing of a model for use in conservation and water quality management. *Freshwater Biology* 36: 451–473.

JMP. 2007. *Statistics and Graphics Guide*. SAS Institute: Cary, North Carolina.

Kay, S. M. and Marple, S. L. 1981. Spectrum analysis: a modern perspective. *Proceedings of the IEEE* 69: 1380–1419.

Khadam, I. B. and Kalvarachchi, J. J. 2006. Water quality modeling under hydrologic variability and parameter uncertainty using erosion-scaled export coefficients. *Journal of Hydrology* 330: 354–367.

Kornecki, T. S., Sabbagh, G. J., and Storm, D. E. 1999. Evaluation of runoff, erosion, and phosphorus modelling system—SIMPLE. *Journal of the American Water Resources Association* 35: 807–820.

Kumar, P. and Foufoula-Georgiou, E. 1997. Wavelet analysis for geophysical applications. *Reviews of Geophysics* 35: 385–412.

Labat, D. 2007. Wavelet analysis of the annual discharge records of the world's largest rivers. *Advances in Water Resources* 31: 109–117.

Lawlor, A. J., Rigg, E., May, L., Woof, C., James, J. B., and Tipping, E. 1998. Dissolved nutrient concentrations and loads in some upland streams of the English Lake District. *Hydrobiologia* 377: 85–93.

Lee, J. S., Hajat, S., Steer, J. P., and Filippi, V. 2008. A time-series analysis of any short-term effects of meteorological and air pollution factors on preterm births in London, UK. *Environmental Research* 106: 185–194.

Lehmann, A. and Rode, M. 2001. Long-tern behavior and cross-correlation water quality analysis of the River Elbe, Germany. *Water Research* 35: 2153–2160.

Li, Y. and Xie, Z. 1997. The wavelet detection of hidden periodicities in time series. *Statistics & Probability Letters* 35: 9–23.

Ljung, G. M. and Box, G. E. P. 1978. On a measure of lack of fit in time series models. *Biometrika* 65: 297–303.

May, L., House, W. A., Bowes, M., and McEvoy, J. 2001. Seasonal export of phosphorus from a lowland watershed: Upper River Cherwell in Oxfordshire, England. *The Science of the Total Environment* 269: 117–130.

Meyer, D. H. and Harris, J. 1991. Prediction of phosphorus loads from non-point sources to South African rivers. *Water South Africa* 17: 211–218.

Moss, B., Johnes, P. J. and Philips, G. 1996. The monitoring of ecological quality and the classification of standing waters in temperate regions: a review and proposal based on a worked scheme for British waters. *Biological Review* 71: 301–339.

Muscutt, A. and Withers, P. 1996. The phosphorus content of rivers in England and Wales. *Water Research* 30: 1258–1268.

Omernik, J. M. 1976. The influence of land use on stream nutrient levels. *USEPA Ecological Research Series* EPA-60013-76-014. USEPA, Corvallis, OR.

Owens, M. 1970. Nutrient balances in rivers. *Proceedings of the Society of Water Treatment Examination* 19: 239–247.

Prairie, Y. T. and Kalff, J. 1986. Effect of watershed size on phosphorus export. *Water Resources Bulletin* 22: 465–469.

Reckhow, K. H, and Simpson, I T 1980. A procedure using modelling and error analysis for the prediction of lake phosphorus concentration from land use information. *Canadian Journal of Fisheries and Aquatic Sciences* 37: 1439–1448.

Rigler, F. H. 1979. The export of phosphorus from Dartmoor watersheds: a model to explain variations of phosphorus concentrations in streamwaters. *Journal of the Marine Biological Association UK* 59: 659–687.

Salcedo, R. I. R., Alvin Ferraz, M. C. M., Alves, C. A., and Martins, F. G. 1999. Time series analysis of air pollution data. *Atmospheric Environment* 33: 2361–2372.

Seeley, R. T. 2006. *An Introduction to Fourier Analysis and Integrals*. Mineola: Dover Publications.

Sharpley, A. N., Kleinman, P. J. A., McDowell, R. W., Gitau, M., and Bryant, R. B. 2002. Modeling phosphorus transport in agricultural watersheds: processes and possibilities. *Journal of Soil and Water Conservation* 57: 425–439.

Simpson, D. M. and De Stefano, A. 2004. A tutorial review of the Fourier transform. *Scope* 13: 28–33.

Sonzogni, W. C. and Lee, G. F. 1972. Nutrient sources for Lake Mendota. Report of the Water Chemistry Program, University of Wisconsin, Madison, p. 49.

Sonzogni, W. C., Chesters, G., Coote, D. R., Jeffs, D. N., Konrad, J. C., Ostry, R. C., and Robinson, J. B. 1980. Pollution from land runoff. *Environmental Science & Technology* 14: 148–153.

Takahashi, D. and Kanada, Y. 2000. High-performance radix-2 and 5 parallel 1-D complex FFT algorithms for distributed-memory parallel computers. *Journal of Supercomputing* 15: 207–228.

Trívez, F. J. 1993. Level shifts, temporary changes and forecasting. *Journal of Forecasting* 14: 543–550.

Ulén, B. 1998. Nutrient exports from two agricultural-dominated watersheds in southern Sweden. *Nordic Hydrology* 29: 41–56.

Vollenweider, R. A. 1968. Scientific fundamentals of the eutrophication of lakes and flowing waters, with particular reference to nitrogen and phosphorus as factors of eutrophication. Rep. No. GP OE/515. Organisation for Economic Co-Operation and Development, Paris, p. 250.

Young, K., Morse, G. K., Scrimshaw, M. D., Kinniburgh, J. H., MacLeod, C. L., and Lester, J. N. 1999. The relation between phosphorus and eutrophication in the Thames watershed, UK. *The Science of the Total Environment* 228: 157–183.

Zellner, A. 2004. *Statistics, Econometrics and Forecasting.* Cambridge: Cambridge University Press.

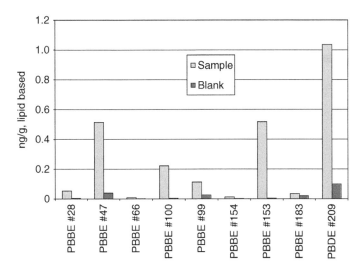

FIGURE 3.3 Comparison of PBDE levels in blank and fish samples. (From Päpke, O., Fürst, P., and Herrmann, T., *Talanta*, 63, 1203–1211, 2004. With permission from Elsevier.)

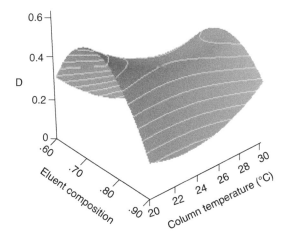

FIGURE 4.22 Response surface image for the main interactive effect of eluent composition/column temperature at predicted critical values with flow rate held constant. (From Gonzales, A., Foster, K. L., and Hanrahan, G., *Journal of Chromatography A*, 1167, 135–142, 2007. With permission from Elsevier.)

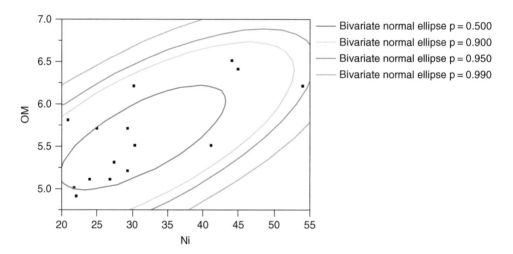

FIGURE 6.3 Density ellipsoid plots for the correlation of OM and NI. This plot includes 50%, 90%, 95%, and 99% confidence curves.

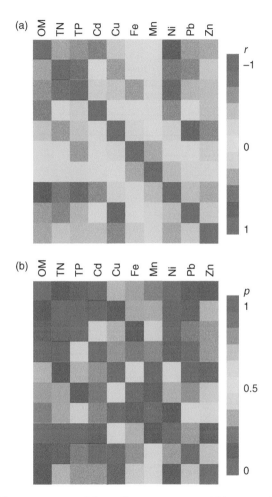

FIGURE 6.6 Color maps among lake sediment parameters: (a) on correlations and (b) on p-values (correlations on a p scale).

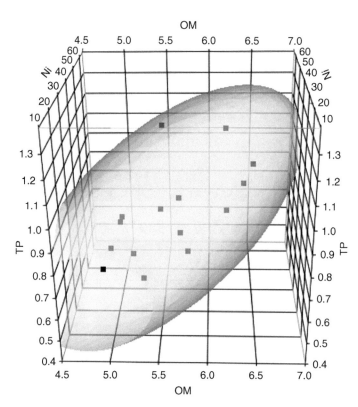

FIGURE 6.7 3D ellipsoid plot of OM, TP, and Ni lake sediment data. A 95% confidence ellipsoid is housed around the data points.

6 Multivariate Data Analysis

6.1 INTRODUCTION TO MULTIVARIATE DATA ANALYSIS

An important objective of environmental monitoring is to understand and characterize chemical and physical measurements of complex systems. The complexity of spatial and temporal dynamics in such systems has spurred interest in the use of multivariate statistical techniques. These methods are appropriate for meaningful data reduction and interpretation of multiparameter investigative efforts. This chapter highlights the use of a myriad of multivariate pattern recognition, multivariate calibration, time series, and correlation techniques and explains their application to real environmental systems.

6.2 DATA PREPROCESSING

Data preprocessing is a very important step in multivariate analysis, which can be used separately (before a method is applied), or as a self-adjusting procedure that forms part of the chemometric approach. Ideally, data preprocessing can be used to remove known interference(s) from data to improve selectivity and enhance more important information to improve robustness techniques such as principal components (see Section 6.4.2), which are scale dependent. If one variable has a much higher variance than the others, it is necessary to scale the original variable before calculating principal components. Common preprocessing scaling techniques include mean centering, autoscaling, column standardization, and autoscaled profiles. Other data preprocessing techniques commonly used in chemometrics include minimum/ maximum transformation, variance normalization, and baseline corrections (1st and 2nd derivative, subtraction).

A closer look at a few of the transformation techniques mentioned above reveals a greater understanding of data preprocessing and applications. Consider the minimum/maximum transformation. Here, the original data matrix $\mathbf{X}$ is converted to $\mathbf{Z}$, a new matrix in which each transformed variable has both a minimum and maximum value of unity:

$$z_{ij} = \frac{x_{ij} - \min(x_j)}{\max(x_j) - \min(x_j)} \tag{6.1}$$

where z_{ij} is a single element in the transformed matrix $\mathbf{Z}$. This technique is often employed for scaling spectral data.

Mean centering is useful when variables experience a similar degree of variance and transformed as

$$z_{ij} = x_{ij} - \bar{x}_j \tag{6.2}$$

Consider a study by Garrido Frenich et al. (1996), for example, that utilized various data preprocessing techniques (including mean centering) to apply multivariate calibration techniques (PLS-1, PLS-2, and PCR) for the simulation high performance liquid chromatography (HPLC) determination of four pyrethroid insecticides (cypermethrin, fenvalerate, *trans*-permethrin, and *cis*-permethrin). Note that multivariate calibration techniques are discussed in detail in Section 6.8. The prediction ability for this study was defined in terms of the root mean square difference (RMSD):

$$\text{RMSD} = \left[\frac{1}{N} \sum_{i-1}^{N} (\bar{x}_i - x_i)^2 \right]^{0.5} \tag{6.3}$$

The effect of the various preprocessing techniques on the RMSD of the calibration matrix for PLS-1, PLS-2, and PCR is presented in Table 6.1. As shown, mean

---

TABLE 6.1

Effect of Preprocessing Techniques on the Relative Prediction Errors of PLS-1, PLS-2, and PCR Models

Model	Preprocessing Technique	Cypermethrin (RSMD)	Fenvalerate (RSMD)	*Trans*-Permethrin (RMSD)	*Cis*-Permethrin (RMSD)
PLS-1	None	0.10	0.21	0.12	0.04
	MC	0.09	0.20	0.12	0.04
	MC + baseline correction	0.09	0.20	0.12	0.04
	MC + scaling	0.78	1.10	0.38	0.13
	MC + smoothing	0.07	0.38	0.14	0.05
PLS-2	None	0.20	0.21	0.12	0.04
	MC	0.18	0.21	0.12	0.04
	MC + baseline correction	0.22	0.20	0.12	0.04
	MC + scaling	0.95	1.48	0.45	0.15
	MC + smoothing	0.18	0.21	0.12	0.04
PCR	None	0.19	0.22	0.13	0.04
	MC	0.19	0.21	0.12	0.04
	MC + baseline correction	0.23	0.20	0.12	0.04
	MC + scaling	0.87	1.98	0.54	0.18
	MC + smoothing	0.19	0.21	0.12	0.04

MC, Mean centering.

Source: Adapted from Garrido Frenich, A., et al., *Journal of Chromatography A*, 727, 27–38, 1996. With permission from Elsevier.

centering had a small but beneficial effect on the dataset presented, ultimately reducing the PLS-1 model dimensionally and RMSD value for cypermethrin, one of the four pyrethroid insecticides. Notice that the PLS-2 and PCR models also resulted in reduced dimensionality and RMSD values in utilizing preprocessed data.

Autoscaling, one of the most popular transformations, is particularly important in environmental analysis where variables monitored are recorded in different units (e.g., pH, concentration, temperature). This can be achieved using the equation

$$z_{ij} = \frac{x_{ij} - \overline{x}_j}{s_{xj}} \tag{6.4}$$

The ability to transform such data into unitless values allows for comparing the distribution among two or more of the monitored variables.

EXAMPLE PROBLEM 6.1

1. Although small in nature, the following water quality dataset is representative of a typical environmental monitoring study involving variables with different units. Use the autoscaling preprocessing technique to arrange data in unitless values for subsequent data analysis.

Sample	Phosphate (mg/L)	Nitrate (mg/L)	Dissolved Oxygen (mg/L)	Conductivity (mS/cm)
1	1.12	7.88	5.56	23.56
2	0.89	9.43	7.55	19.66
3	0.91	7.89	3.41	28.65
4	1.24	4.16	4.39	44.32
5	1.33	10.43	6.98	35.78
6	0.89	6.68	4.67	19.98
7	0.60	9.90	5.00	23.56
8	0.44	6.13	4.32	44.67
Mean	0.93	7.85	5.24	30.02
Standard deviation	0.30	2.08	1.40	10.32

Answer: Using Equation 6.3, transformation results in the following autoscaled data matrix:

Sample	Phosphate (mg/L)	Nitrate (mg/L)	Dissolved Oxygen (mg/L)	Conductivity (mS/cm)
1	0.63	0.01	0.23	−0.62
2	−0.13	0.76	1.65	−1.00
3	−0.06	0.02	−1.30	−0.13
4	1.03	−1.77	−0.61	1.38
5	1.33	1.24	1.24	0.56
6	−1.33	−0.56	−0.41	−0.98
7	−1.1	0.99	−0.17	−0.63
8	−1.63	−0.68	−0.66	1.42

6.3   CORRELATION ANALYSIS

As stated in Chapter 2, the aim of correlation analysis is to detect the relationships among variables with the measure of linear correlation assessed by the correlation coefficient. The importance of bivariate correlation has been fully discussed but multivariate datasets still need attention. Here, the proportion of dependent variables (Y) that can be attributed to the combined effects of all the X independent variables acting together are considered. A number of correlation techniques and visual tools can be used to assess relationships within multivariate datasets. Such techniques are discussed in detail here using a generic multivariate lake sediment dataset highlighting common measurable parameters (Table 6.2).

6.3.1   CORRELATION MATRIX

A matrix of *correlation coefficients* that summarizes the strength of the linear relationships from the sediment data is presented in Table 6.3. Typically, coefficients from ±0.8 to 1.0 indicate a strong correlation, while values from ±0.5 to 0.8 indicate a moderate correlation, and values less than ±0.5 indicate a weak correlation (Hupp et al. 2008). Organic matter (OM) showed weak to moderate correlation with Pb (0.43), Cd (0.54), and Ni (0.71), confirming a suspicion of higher OM levels retaining heavy metals. Intermetallic relationships as shown by Cd/Ni (0.50), Zn/Cu (0.50), and Cu/Pb (0.60) reveal a similar behavior during transport in the aquatic system. Additionally, total nitrogen (TN) showed a moderate correlation with total phosphorus (TP) (0.60).

Consider a study by Chatterjee et al. (2007) who examined the distribution and possible sources of trace elements in the sediment cores of a tropical macrotidal estuary. Although performed on a different type of aquatic system, correlation analysis (*correlation matrix*) results of this study related relatively well with the simulated sediment data

---

TABLE 6.2
Descriptive Statistics for a Generic Lake Sediment Dataset

Parameter	Mean ($n = 15$)	Standard Deviation
Organic matter (OM)[a]	2.21	0.17
Total nitrogen (TN)[b]	1.56	0.41
Total phosphorus (TP)[b]	0.72	0.12
Cd[c]	12.1	1.00
Cu[c]	32.2	2.58
Fe[c]	7879	162
Mn[c]	724	47.1
Ni[c]	27.8	7.33
Pb[c]	75.1	15.6
Zn[c]	96.8	4.72

[a] Units = %
[b] Units = mg g^{-1} dry weight
[c] Units = μg g^{-1} dry weight

TABLE 6.3

Correlation Matrix for the Generic Lake Sediment Dataset

	OM	TN	TP	Cd	Cu	Fe	Mn	Ni	Pb	Zn
OM	1.00	0.40	0.46	0.54	0.27	0.08	−0.05	0.71	0.44	0.39
TN	0.39	1.00	0.60	0.00	0.44	−0.08	0.09	0.57	0.50	0.27
TP	0.46	0.60	1.00	0.22	0.27	0.44	−0.15	0.63	0.32	0.30
Cd	0.54	0.00	0.22	1.00	0.07	0.04	−0.06	0.50	−0.02	0.30
Cu	0.26	0.44	0.27	0.07	1.00	0.17	0.01	0.20	0.65	0.50
Fe	0.08	−0.08	0.44	0.04	0.17	1.00	0.38	−0.09	−0.07	0.25
Mn	−0.05	0.09	−0.15	−0.06	0.01	0.38	1.00	−0.41	−0.19	0.15
Ni	0.71	0.57	0.63	0.50	0.20	−0.09	−0.41	1.00	0.29	0.39
Pb	0.43	0.50	0.32	−0.02	0.66	−0.07	−0.20	0.29	1.00	0.25
Zn	0.39	0.27	0.30	0.30	0.50	0.25	0.15	0.39	0.25	1.00

presented in Table 6.2. Significant correlations of organic carbon (OC) with most of the elements (excluding B, Ba, and Mn) were observed, which further strengthens the argument that sediment organic matter plays an important role in element distributions. Strong intermetallic relationships were also evident. Note, however, that the correlation matrix is limited to the extent that it only incorporates observations that have nonmissing values for all parameters in the analysis. Consider further the following techniques.

6.3.2 INVERSE CORRELATIONS, PARTIAL CORRELATIONS, AND COVARIANCE MATRIX

The inverse correlation matrix is shown in Table 6.4. The diagonal elements of the matrix, often termed the variance inflation factors (VIF), are a function of how

TABLE 6.4

Inverse Correlation Matrix for the Generic Lake Sediment Dataset

	OM	TN	TP	Cd	Cu	Fe	Mn	Ni	Pb	Zn
OM	4.06	1.73	−0.09	−0.59	0.27	−0.01	−2.09	−3.93	−2.19	0.42
TN	1.73	8.27	−4.61	1.81	−2.32	3.97	−5.38	−5.36	−1.41	1.42
TP	−0.09	−4.61	6.22	−0.77	1.63	−4.37	2.81	−0.24	−0.35	−0.36
Cd	−0.59	1.81	−0.77	2.16	−0.78	0.81	−1.05	−1.58	0.43	0.20
Cu	0.27	−2.32	1.63	−0.78	3.18	−1.66	1.36	1.13	−1.46	−1.21
Fe	−0.01	3.97	−4.37	0.81	−1.66	4.47	−2.76	−0.43	0.38	0.24
Mn	−2.09	−5.38	2.81	−1.05	1.36	−2.76	5.31	4.99	1.54	−1.43
Ni	−3.93	−5.36	−0.24	−1.58	1.13	−0.43	4.99	9.62	2.38	−2.06
Pb	−2.19	−1.41	−0.35	0.43	−1.46	0.38	1.54	2.38	3.44	−0.17
Zn	0.42	1.42	−0.36	0.20	−1.21	0.24	−1.43	−2.06	−0.17	2.13

closely a given parameter is a linear function of the other parameters. The inverse correlation coefficient is denoted R^{-1}. The diagonal element is denoted r^{ii} and is computed as

$$r^{ii} = VIF_i \frac{1}{1 - R_i^2} \tag{6.5}$$

where R_i^2 is the coefficient of variation from the model regressing the ith explanatory parameter on the other explanatory parameters. Given this, a large r^{ii} thus indicates that the ith variable is highly correlated with any number of the other variables.

The partial correlation matrix from the sediment dataset is presented in Table 6.5. This table shows the partial correlations of each pair of parameters after adjusting for all other parameters. Note that this is simply the negative of the inverse correlation matrix presented in Table 6.4, scaled to unit diagonal.

The *covariance matrix* (Table 6.6) displays the covariance matrix for the analysis of the sediment dataset. The covariance of two datasets can be defined as their tendency to vary together. Note that both the covariance and correlation matrices must be calculated for the application of factor analytic methods (see Section 6.4.2).

6.3.3 Pairwise Correlations

The pairwise correlations from the sediment data is presented in Table 6.7 and lists the Pearson product-moment correlations (PPMC) for each pair of Y variables using all available values and is calculated as

$$r = \frac{\sqrt{\Sigma (x - \bar{x})(y - \bar{y})}}{\sqrt{\Sigma (x - \bar{x})^2} \sqrt{\Sigma (y - \bar{y})^2}} \tag{6.6}$$

TABLE 6.5

Partial Correlation Matrix for the Generic Lake Sediment Dataset

	OM	TN	TP	Cd	Cu	Fe	Mn	Ni	Pb	Zn
OM	.	−0.29	0.01	0.19	−0.07	0.00	0.45	0.62	0.58	−0.14
TN	−0.29	.	0.64	−0.42	0.45	−0.65	0.81	0.60	0.26	−0.34
TP	0.01	0.64	.	0.21	−0.36	0.82	−0.48	0.03	0.07	0.09
Cd	0.19	−0.42	0.21	.	0.29	−0.26	0.31	0.34	−0.16	−0.09
Cu	−0.07	0.45	−0.36	0.29	.	0.44	−0.33	−0.20	0.44	0.46
Fe	0.00	−0.65	0.82	−0.26	0.44	.	0.56	0.06	−0.09	−0.08
Mn	0.45	0.81	−0.48	0.31	−0.33	0.56	.	−0.69	−0.36	0.42
Ni	0.62	0.60	0.03	0.34	−0.20	0.06	−0.69	.	−0.41	0.45
Pb	0.58	0.26	0.07	−0.16	0.44	−0.09	−0.36	−0.41	.	0.06
Zn	−0.14	−0.34	0.09	−0.09	0.46	−0.08	0.42	0.45	0.06	.

TABLE 6.6
Covariance Matrix for the Generic Lake Sediment Dataset

	OM	TN	TP	Cd	Cu	Fe	Mn	Ni	Pb	Zn
OM	0.27	0.12	0.03	0.28	0.36	8.34	−1.68	3.67	3.56	0.97
TN	0.12	0.37	0.05	0.00	0.69	−10.5	3.47	3.42	4.76	0.76
TP	0.03	0.05	0.02	0.03	0.10	13.3	−1.44	0.98	0.79	0.22
Cd	0.28	0.00	0.03	1.00	0.18	8.63	−4.00	4.92	−0.45	1.42
Cu	0.36	0.69	0.10	0.18	6.67	87.4	1.94	5.18	26.3	6.14
Fe	8.34	−10.5	13.38	8.63	87.4	38407	4591	−190	−218	229
Mn	−1.68	3.47	−1.44	−4.00	1.94	4591	3863	−250	−193	45.4
Ni	3.67	3.42	0.98	4.92	5.18	−190	−250	98.7	45.17	18.3
Pb	3.58	4.76	0.79	−0.45	26.3	−218	−193	45.1	243	18.5
Zn	0.97	0.76	0.22	1.42	6.14	229	45.4	18.3	18.5	22.3

TABLE 6.7
Pairwise Correlation Report Including Significance Probabilities

Parameter 1	Parameter 2	Correlation	Count	Probability
TN	OM	0.39	15	0.1431
TP	OM	0.46	15	0.0827
TP	TN	0.59	15	0.0192*
Cd	OM	0.54	15	0.0376*
Cd	TN	0.00	15	0.9738
Cd	TP	0.21	15	0.4397
Cu	OM	0.26	15	0.3375
Cu	TN	0.43	15	0.1013
Cu	TP	0.26	15	0.3378
Cu	Cd	0.07	15	0.7972
Fe	OM	0.08	15	0.7735
Fe	TN	−0.08	15	0.7540
Fe	TP	0.43	15	0.1046
Fe	Cd	0.04	15	0.8762
Fe	Cu	0.17	15	0.5383
Mn	OM	−0.05	15	0.8549
Mn	TN	0.09	15	0.7449
Mn	TP	−0.14	15	0.5973
Mn	Cd	−0.06	15	0.8197
Mn	Cu	0.01	15	0.9659
Mn	Fe	0.37	15	0.1661
Ni	OM	0.70	15	0.0033*
Ni	TN	0.56	15	0.0278*

continued

TABLE 6.7 (continued)

Parameter 1	Parameter 2	Correlation	Count	Probability
Ni	TP	0.63	15	0.0116*
Ni	Cd	0.49	15	0.0605
Ni	Cu	0.20	15	0.4701
Ni	Fe	−0.09	15	0.7284
Ni	Mn	−0.40	15	0.1339
Pb	OM	0.43	15	0.1022
Pb	TN	0.50	15	0.0570
Pb	TP	0.32	15	0.2380
Pb	Cd	−0.02	15	0.9178
Pb	Cu	0.65	15	0.0081*
Pb	Fe	−0.07	15	0.8001
Pb	Mn	−0.19	15	0.4767
Pb	Ni	0.29	15	0.2920
Zn	OM	0.39	15	0.1452
Zn	TN	0.26	15	0.3355
Zn	TP	0.30	15	0.2749
Zn	Cd	0.30	15	0.2761
Zn	Cu	0.50	15	0.0558
Zn	Fe	0.24	15	0.3736
Zn	Mn	0.15	15	0.5822
Zn	Ni	0.39	15	0.1487
Zn	Pb	0.25	15	0.3649

* Statistically significant.

The pairwise correlation report also lists the count values and significant probabilities, and compares the correlations with a bar chart (Figure 6.1). PPMC allows the pairwise comparison of samples, and therefore, is independent of the total number of samples in the analysis. Thus, this approach is especially useful for environmental applications. Note in Table 6.7 that the $p < 0.05$ are considered statistically significant and are labeled with an asterisk.

6.3.4 FIT FUNCTIONS

Fit functions allow investigators to explore how the distribution of one continuous variable relates to another. The analysis begins as a scatterplot of points, to which one can interactively add other types of fitting functions including:

1. simple *linear regression*;
2. polynomial regression of selected degree;
3. smoothing spline;

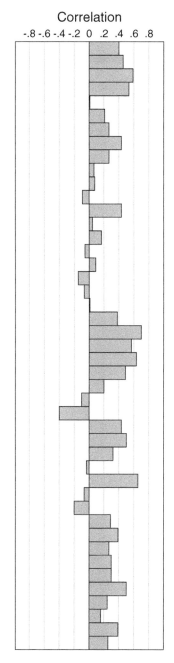

FIGURE 6.1 Bar chart of pairwise correlations on the generic lake sediment data.

4. bivariate normal density ellipses;
5. bivariate nonparametric density contours;
6. multiple fit over groups.

Consider the strong pairwise correlation (0.70) between OM and Ni in the exploration of a few of the fit functions. Figure 6.2 presents simple linear regression (Figure 6.2a), polynomial quadratic regression (Figure 6.2b), and polynomial cubic regression (Figure 6.2c) of the two parameters including 95% confidence limits around the data points. If the confidence area of the regression line includes the response mean, the slope of the line of fit is not significantly different from zero at the 0.05 significance level. An r^2 value is reported and measures the proportion of the variation around the mean explained by the simple linear or polynomial model. The remaining variation is attributed to random error.

Now consider the *density ellipse* function that produces an ellipse containing the data points and is computed from the bivariate normal distribution fit to the X and Y variables. The bivariate normal density is a function of the means and standard deviations of the X and Y variables and the correlation between them. Figure 6.3 displays the density ellipsoid for the correlation of OM and Ni and includes 50%, 90%, 95%, and 99% confidence curves. Note that the ellipsoids are both density contours and confidence curves. As confidence curves, they show where a given percentage of the data is expected to lie (assuming a bivariate normal distribution). For interpretation, the ellipsoid collapses diagonally as the correlation between the two parameters approaches 1 or -1.

6.3.5   SCATTERPLOT MATRIX

The *scatterplot matrix* is a visual tool in which to check the pairwise correlations between given parameters. Figure 6.4 contains all of the pairwise scatterplots of the sediment parameters on a single page in a matrix format. Notice that 95% bivariate normal density ellipses are imposed on each scatterplot. If the variables are bivariate normally distributed, the ellipses enclose approximately 95% of the points. The correlation of the variables is seen by the collapsing of the ellipse along the diagonal axis. If the ellipse is fairly round and is not diagonally oriented, the variables are uncorrelated.

Sometimes it can be helpful to overlay other features (e.g., histogram, fitted curve) for interpretation purposes. Consider a closer look at TN and TP with the scatterplot displayed in Figure 6.5. Both the histogram and the correlation coefficient have been added to the scatterplot for greater interpretation.

6.3.6   COLOR MAPS AND 3D ELLIPSOID PLOTS

Generated *color maps* are also useful tools in visualizing correlations among parameters. Normally, color maps on correlations are produced as shown in Figure 6.6a, using the sediment data. The correlations among parameters are presented on a scale from +1 to 0 to −1. A color map of *p*-values (Figure 6.6b) shows the significance of the correlations on a *p*-scale from 1 to 0.5 to 0.

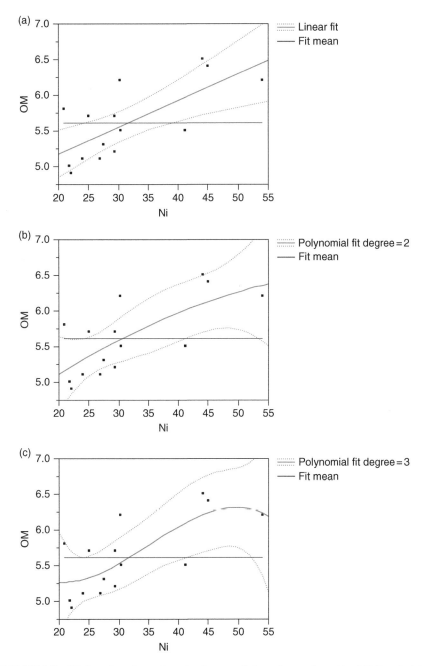

FIGURE 6.2 Fit functions from the pairwise correlation of organic matter (OM) and nickel (Ni): (a) simple linear regression, (b) polynomial quadratic regression, and (c) polynomial cubic regression. Note that 95% confidence limits around the data points. The fit mean is represented by the solid horizontal line.

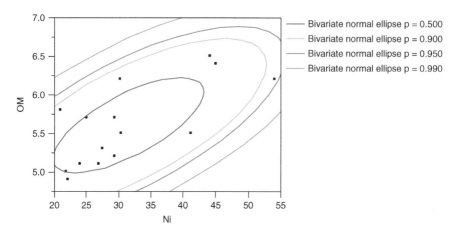

FIGURE 6.3   (See color insert following page 206.) Density ellipsoid plots for the correlation of OM and Ni. This plot includes 50%, 90%, 95%, and 99% confidence curves.

A 3D ellipsoid plot is another unique tool in visualization and can be used to gain a better understanding of three chosen parameters. Figure 6.7, for example, displays a 3D ellipsoid plot of OM, TP, and Ni sediment data and includes a 95% confidence ellipsoid around the data points.

6.3.7 NONPARAMETRIC CORRELATIONS

Section 2.3 touches upon the use of nonparametric methods in testing the relationships among variables. Closer examination here reveals the statistical concepts behind them as well as example applications in which they are utilized.

Spearman's rho, often termed Spearman's rank correlation coefficient, differs from the PPMC in that it computes correlation based on the rank of the data values, and not on the values themselves. It is advantageous as it is a scale-free procedure where means, variances and monotonicity of distances among ordered pairs would have no effect on the results of an analysis. It does, however, use the PPMC formula after ranking is achieved. When converting to ranks, the smallest value on X becomes a rank of 1, the second value a rank of 2, and so on. Meklin et al. (2007) utilized the Spearman's rho test to test the correlation among 36 mold species in indoor and outdoor air samples. The measure of both indoor and outdoor samples is a common technique to determine if a building has an abnormal mold condition. However, only four species were found to be correlated in both indoor and outdoor samples, suggesting that evaluating the mold burden indoors by a simple genus-level comparison may be misleading and thus further investigation is warranted.

Spearman's rank correlation is satisfactory for testing a null hypothesis of independence between two variables but it is difficult to interpret when the null hypothesis is rejected (Rupinski and Dunlap 1996). In such cases, *Kendall's tau* is useful. It measures the degree of correspondence between two rankings and assesses

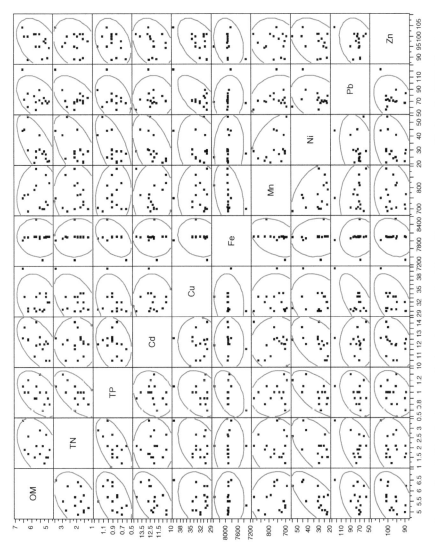

FIGURE 6.4 Scatter plot matrix of all pairwise correlations between the lake sediment parameters. Notice the 95% bivariate normal density ellipse imposed on each scatter plot.

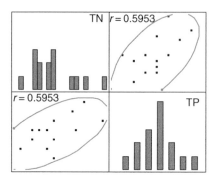

FIGURE 6.5 Individual scatter plots of TN and TP with an embedded histogram and correlation coefficient.

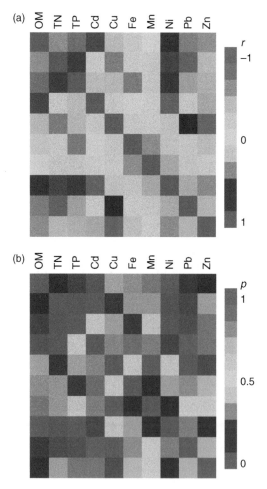

FIGURE 6.6 **(See color insert following page 206.)** Color maps among lake sediment parameters: (a) on correlations and (b) on p-values (correlations on a p-scale).

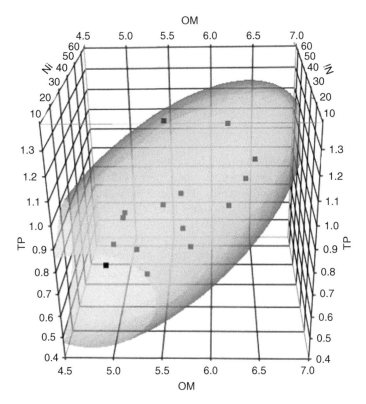

FIGURE 6.7 (**See color insert following page 206.**) 3D ellipsoid plot of OM, TP, and Ni lake sediment data. A 95% confidence ellipsoid is housed around the data points.

the significance of this correspondence, based on the number of *concordant* and *discordant* pairs of observations. Kendall's tau is calculated as (JMP 2007)

$$\tau_b = \frac{\sum_{i \le j} \text{sgn}(x_i - x_j)\,\text{sgn}(y_i - y_j)}{\sqrt{(T_0 - T_1)(T_0 - T_2)}} \tag{6.7}$$

where $T_0 = [n(n-1)]/2$, $T_1 = \Sigma[t(t_i)(t_i - 1)]/2$, and $T_2 = \Sigma[(u_i)(u_i - 1)]/2$. Note that the t_i (u_i) are the number of tied x (respectively y) values in the ith group of tied x (respectively y) values, n is the number of observations, and Kendall's τ_b ranges from -1 to $+1$ (JMP 2007).

Laranjeira et al. (2000) utilized Kendall's τ_b to assess environmental tobacco smoke exposure (using expired carbon monoxide levels) in nonsmoking waiters before and after a normal day's shift and to compare pre-exposure levels with nonsmoking medical students. Results show no correlation between reported levels of exposure in the workplace and postexposure carbon monoxide levels (Kendall's $\tau_b = 0.07$, $p > 0.2$) but a strong correlation between the number of tables available for smokers and postexposure carbon monoxide levels was found (Kendall's $\tau_b = 0.2$, $p < 0.0001$).

A pair is *concordant* if the observation with the larger value of X also has the largest Y. They are *discordant* if the observation with the largest X has the smaller value of Y.

Another nonparametric correlation procedure, *Hoeffding's D*, can be used to test association. In this test, D values range from -0.5 to 1.0, with 1.0 representing complete dependence. In other words, the higher the value of D, the more dependent are X and Y. The D used here is 30 times Hoeffding's original D (Hoeffding 1948). The formula is (JMP 2007)

$$D = 30\left(\frac{(n-2)(n-3)D_1 + D_2 - 2(n-2)D_3}{n(n-1)(n-2)(n-3)(n-4)}\right) \qquad (6.8)$$

where $D_1 = S_i(Q_i - 1)(Q_i - 2)$, $D_2 = S_i(R_i - 1)(S_i - 1)(S_i - 2)$, and $D_3 = (R_i - 1)(S_i - 2)(Q_i - 1)$. Note that R_i and S_i are ranks of the X and Y values, and the Q_i are one plus the number of points that have both X and Y values less than the ith points (JMP 2007).

6.4 PATTERN RECOGNITION—UNSUPERVISED

Pattern recognition techniques seek to identify similarities and regularities present in a given dataset to attain natural classification or groupings. Two general categories are considered: unsupervised and supervised methods. In unsupervised methods, information on the individual groupings is likely unknown but not necessarily considered a prerequisite for data analysis. Consider the following techniques.

6.4.1 CLUSTER ANALYSIS

Cluster analysis (CA) includes a number of powerful algorithms and methods for grouping objects of similar kinds into organized categories. More specifically, CA is an exploratory analysis tool that aims to sort different objects into groups in such a way that the degree of association between two objects is maximal if they belong to the same group and minimal if not (Gong and Richman 1995). The first step in CA is the establishment of the similarity or distance. Most clustering algorithms partition the data based on similarity; the more similar, the more likely that they belong to the same cluster. *Euclidean distance*, the square root of the sum of the square of the x difference plus the square of the y distance, is the most common distance measure. The three most widely used clustering methods are hierarchical, k-means, and normal mixtures.

Hierarchical clustering entails the use of agglomerative methods (establishment of a series of n objects into groups) and divisive methods (further separation of objects into finer groupings) that finds clusters of observations within a dataset. Agglomerative techniques are more commonly used and hence are covered in this book. Combining of groups into hierarchical clustering is visually represented by a two-dimensional *dendogram*—a tree diagram that lists each observation, and shows which cluster it is in and when it entered its cluster. Five widely used algorithms employed in hierarchical clustering include average linkage, single linkage, complete linkage, centroid linkage, and Ward's linkage. These linkage algorithms are based on measurements of proximity between two groups of objects. The term n_x is the number of objects in cluster x and n_y is the number of objects in cluster y, and x_{xi} is the *i*th object in cluster x. Calculations are as follows:

1. *Average linkage*—uses the distance between two clusters (defined as the average of distances between all pairs of observations):

$$d(x,y) = \frac{1}{n_x n_y} \sum_{i=1}^{n_x} \sum_{j=1}^{n_y} \text{dist}(x_{xi}, x_{yj}) \qquad (6.9)$$

2. *Single linkage* (also termed nearest neighbor)—uses the smallest distance between objects in the two groups:

$$d(x, y) = \min(\text{dist}(x_{xi}, x_{yj})), \quad i \in (i, \dots, n_x), \quad j \in (1, \dots, n_y) \qquad (6.10)$$

3. *Complete linkage* (also termed furthest neighbor)—uses the largest distance between objects in two groups:

$$d(x, y) = \max(\text{dist}(x_{xi}, x_{yj})), \quad i \in (i, \dots, n_x), \quad j \in (1, \dots, n_y) \qquad (6.11)$$

4. *Centroid linkage*—uses the distance between the centroids of the two groups:

$$d(x, y) - d(\bar{x}_x, \bar{x}_y) \qquad (6.12)$$

where $\bar{x}_x = \dfrac{1}{n_x} \sum_{i=1}^{n_x} x_{xi}$. Note that $\bar{x}_y$ is defined similarly.

5. *Ward's linkage*—uses ANOVA sum of squares; the increase in the total within-group sum of squares as a result of joining groups x and y

$$d(x, y) = n_x n_y \, d_{xy}^2 / (n_x + n_y) \qquad (6.13)$$

where d_{xy}^2 is the distance between cluster x and y defined in the centroid linkage.

So how do investigators choose the appropriate method? The average linkage method is more appropriate to join clusters with smaller variances when the research purpose is homogeneity within clusters (Fraley and Raftery 2002). It also has the advantage of leading directly to the best dendogram. Single linkage has been shown to have numerous desirable properties including simplicity, but to fare poorly if outliers are noticeably present (Kopp 1978). Complete linkage can also be distorted by outliers but tends to form dendograms with more structure (and is hence more informative) than the single linkage method (Milligan 1980). Centroid linkage has been shown to be more robust in the presence of outliers, but are reported to form less effectively than the average linkage method (Milligan 1980). Although also sensitive to outliers, the Ward's linkage method is arguably the most efficient and widely used clustering algorithm. It is distinct from the others in that it uses an analysis of variance approach to evaluate the distances among clusters and attempts to minimize the sum of squares of any two clusters that can form at each step of the process (Milligan 1980, Bock 1985).

Consider CA examination of the generic lake sediment data presented in Table 6.2. Hierarchical CA using the Ward's method can prove fruitful as is evident from the dendogram presented in Figure 6.8. The corresponding cluster history and Euclidean distances are provided in Table 6.8. Two cluster sets are visibly apparent, cluster 1 (OM, Ni, Pb, Mn, Cd, Cu) and cluster 2 (TN, TP, Zn, Fe). TN and TP relationships in sediments have been well established (Koehler et al. 2005, Trevisan and Forsberg 2007). The TP sediment pool, for example, is composed of a number of operationally defined phosphorus components (Chapter 1), each with their own release pathway affected by equilibrium conditions between the surface sediment and the overlaying water (Amini and McKelvie 2004, Spears et al. 2007). Moreover, fluctuations in TP have been explained by variations in total Fe and dissolution of carbonate associated phosphate (Knöscke 2005).

K-means clustering employs an algorithm to cluster *n* objects based on attributes into *k* partitions, where $k < n$, and *k* is a positive integer that needs to be determined prior to the analysis. There are many variants to this algorithm that has a simple iterative solution for finding a locally minimal solution. The more commonly used

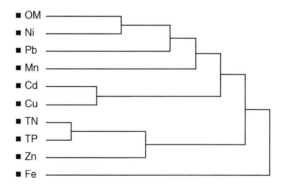

FIGURE 6.8 Dendogram of lake sediment data produced by CA employing the Ward's linkage method.

TABLE 6.8

Clustering History of the Nine Clusters with Corresponding Linkage Distances for Surface Sediment Parameters

Cluster	Distance	Leader	Joiner
9	0.23355297	TN	TP
8	0.23463316	Cd	Cu
7	0.23540554	OM	Ni
6	0.40468806	TN	Zn
5	0.44612433	OM	Pb
4	1.07182915	OM	Mn
3	1.57160539	OM	Cd
2	2.28437277	OM	TN
1	11.6032351	OM	Fe

Lloyd's algorithm has an objective to achieve minimized total intracluster variance, or, the squared error function (Hartigan and Wong 1979):

$$V = \sum_{i=1}^{k} \sum_{x_j \in S_i} (x_j - \mu_i)^2 \qquad (6.14)$$

where $S_i = k$ clusters with $i = 1, 2, \ldots, k$ and μ_i is the centroid or mean point of all the points. This algorithm consists of a simple re-estimation procedure. The centroid is computed for each set and every point is assigned to the cluster whose centroid is closest to that point. These two steps are alternated until a stopping criterion is met (i.e., when there is no further change in the assignment of the data points). One potential criticism of the k-means algorithm is that the number of clusters (k) needs to be investigator assigned before analysis. If an inappropriate choice of k is made, less than agreeable results can be obtained. Since it is a relatively fast algorithm, this method can be run numerous times to help return the best clustering scenario.

6.4.2 FACTOR ANALYTIC TECHNIQUES

Factor analytic techniques such as *factor analysis* (FA) and *principal component analysis* (PCA) are both variable reduction techniques used to uncover the dimensions of a set of variables. These two powerful techniques, however, are often confused as the same statistical technique, with many differences existing between them and in the types of analyses to which they are each best suited (Table 6.9). In order to better understand these differences and gain greater insight to factor analytic methods, key concepts and terms must be defined:

1. *Exploratory factor analysis* (EFA)—explores the underlying structure of a large set of variables when there is no *a priori* knowledge of the factor structure.

TABLE 6.9

Basic Underlying Differences between Factor Analysis and Principal Component Analysis

Factor Analysis	Principal Component Analysis
Preferred technique for structure detection.	Preferred technique for data reduction.
Attempts to separate the common variance affected by more than one variable.	Uses whole variability in the analysis.
Identifies the number of latent constructs and the underlying factor structure of a set of variables.	Reduces the number of observed variables to a smaller number of principal components.
Offers *a priori* knowledge to help solve relevant problems.	A mathematical manipulation to recast variables and factors.
Observed values are linear combinations of underlying and unique factors.	*Scores* are a linear combination of the observed variables weighted by eigenvectors.

2. *Confirmatory factor analysis* (CFA)—explores the relationships among a smaller set of variables when there is advanced knowledge of what these relationships might look like. Accuracy of such hypotheses can then be tested.

3. *Factors* (from FA)—represent the common variance of variables (excluding unique variance) and are highly correlation intensive.

4. *Components* (from PCA)—represent both common and unique variance of interested variables seeking to reproduce both the total variable variance with all components and to reproduce the correlations.

5. *Factor loadings* (component loadings in PCA)—are the correlation coefficients between the variables and factors.

6. *Factors scores* (component scores in PCA)—linear combinations of variables that are used to estimate the case's scores on the factors or components. To compute the factor score for a given case for a given factor, an investigator takes the case's standardized score on each variable, multiplies by the corresponding factor loading of the variable for the given factor, and sums these products.

7. *Factor rotation*—a process that transforms the values of the factor loadings to meet the requirements specified by investigators;

8. *Commonality*—the variance in observed variables accounted for by a common factor.

The underlying principle of both PCA and FA is that chosen variables can be transformed into linear combinations of an underlying set of hypothesized or unobserved factors (Conway and Huffcutt 2003). Factors may then be either associated with two or more of the original variables (common factors) or associated with an individual variable (unique factors). Factor loadings discussed earlier relate the association between factors and original variables. Therefore, in factor analytic techniques it is necessary to find the loadings, and then solve for the factors, which will approximate the relationship between the original variables and underlying factors. To analyze data with factor analytic techniques, four key steps must be logically performed:

1. data collection, preprocessing, and formation of a data matrix;
2. the use of an appropriate factor extraction method;
3. selection of a proper factor rotation method;
4. interpretation and construction of factor scales.

So how do investigators know which method to use in a given analysis? If the purpose is to identify the latent variables contributing to the common variance in a set of measured variables, FA is used. Here, only the variability in a factor that has commonality with other factors is used in the analysis. Often, it is difficult to interpret loadings. *Factor analysis* is preferred when the goal of the analysis is to detect structure.

If the purpose of the analysis is to reduce the information in a large number of variables into a set of weighted linear combinations, PCA is the appropriate method. It is a procedure for finding hypothetical variables (components), which account for as much of the variance in multidimensional datasets as possible. *Principal component analysis* is arguably the most widely used of the two techniques in environmental analyses and is therefore the main feature of this chapter. Discussion on the four factor analytic steps listed earlier is incorporated into the PCA example for greater understanding.

Principal components (PCs) should be thought of as vehicles to map the structure of multivariate data sets as completely as possible using as few variables as possible. Ostensibly, it reduces it to dimensionalities that are graphable for greater interpretation. Each PC is calculated by taking a linear combination of an eigenvector of the correlation matrix with an original variable. The PCs are further derived from eigenvalue decomposition of the correlation matrix, the covariance matrix or from the unscaled and uncentered data. Computation of eigenvalues and eigenvectors via the correlation matrix is outlined below. Given an identity matrix ($\mathbf{I}$), a zero vector ($\mathbf{0}$) and a correlation matrix ($\mathbf{R}$) (Mason and Young 2005)

$$\mathbf{I} = \begin{bmatrix} 1 & 0 \\ 0 & 1 \end{bmatrix}, \qquad \mathbf{0} = \begin{bmatrix} 0 \\ 0 \end{bmatrix}, \qquad \text{and} \qquad \mathbf{R} = \begin{bmatrix} 1 & r \\ r & 1 \end{bmatrix},$$

where r is the correlation between two variables. Values of (λ_1, $\mathbf{a}_1$) and (λ_2, $\mathbf{a}_2$) that satisfy the matrix equation

$$(\mathbf{R} - \lambda_i \mathbf{I})\mathbf{a}_i = 0$$

for $i = 1, 2$ are sought. The terms λ_1 and λ_2 are eigenvalues of $\mathbf{R}$, with $\mathbf{a}_1$ and $\mathbf{a}_2$ the eigenvectors of $\mathbf{R}$, where

$$\mathbf{a}_1 = \begin{bmatrix} a_{11} \\ a_{12} \end{bmatrix}, \qquad \text{and} \qquad \mathbf{a}_2 = \begin{bmatrix} a_{11} \\ a_{12} \end{bmatrix}$$

For *n* original variables, *n* principal components are constructed.

1. The first PC is the linear combination of the original variables that has the greatest possible variance.
2. The second PC is a linear combination of both variables that is uncorrelated with the first PC and has the second largest amount of variation.

In PCA, the original data matrix, $\mathbf{X}$, is approximated by the product of two small matrices—the score and loading matrices:

$$\mathbf{X} = \mathbf{TL}^{\mathrm{T}} \tag{6.15}$$

where $\mathbf{X}$ is the original data matrix, $\mathbf{T}$ the scores matrix, and $\mathbf{L}$ is the loading matrix. The factor loadings, also called component loadings in PCA, are the correlation coefficients between the variables (rows) and factors (columns). To obtain the percent of variance in all the variables accounted for by each factor, the sum of the squared factor loadings for that factor (column) is taken and divided by the number of variables present in the study. Principal component scores, also termed component scores in PCA, are the scores of each case (row) on each factor (column). To compute the factor score for a given case for a given factor, investigators take the standardized score on each variable, multiply by the corresponding factor loading of the variable for the given factor, and finally total these products. Note that PCs are dependent on the units used to measure the original variables as well as the range of values they assume. Thus, standardization via data preprocessing should be done prior to performing PCA. Recall the discussion in Section 6.2. A variety of techniques are used to estimate the number of PCs including percentage of explained variance, eigenvalue-one criterion, *scree-test*, and *cross validation* (Otto 1999). If all possible PCs are used in the model, then 100% of the variance is explained. Note, however, the use of all PCs is not justified in routine analysis. In PCA, the ultimate question is, how many factors do we want to extract? Also note that as the extraction of consecutive factors occurs, they account for less and less overall variability. The decision of when to stop extracting factors is arguable, but depends primarily on when there is only very little random variability remaining.

Consider the lake water physicochemical dataset presented in Table 6.10. Mean parameters ($n = 15$ samples) were analyzed using the Kolmogorov–Smirnov test ($p > 0.05$). With the exception of TP and total dissolved solids (TDS), water quality parameters exhibited a normal distribution. Principal component analysis can be utilized to elucidate the physicochemical makeup of this aquatic environment, and be instructive in distinguishing between natural and anthropogenic sources of pollution in such systems. Therefore, PCA was performed on the correlation matrix to gain a more reliable display method and greater understanding about the relationships within the dataset.

The resultant PCA report is shown in Table 6.11. As shown, the cumulative contribution of the first four PCs account for roughly 71% of the cumulative variance in the lake water dataset. The corresponding scree plot (Figure 6.9) shows the sorted eigenvalues as a function of the eigenvalue index. After the curve starts to

TABLE 6.10
Physicochemical Water Quality Parameters for PCA Analysis

Parameter	Mean ($n = 15$)	Standard Deviation
pH	9.42	0.051
Dissolved oxygen (DO)[a]	8.71	0.171
Conductivity[b]	45.7	0.220
Temperature[c]	25.7	0.192
Total dissolved solids (TDS)[a]	690	9.441
Total nitrogen (TN)[a]	0.98	0.024
Total phosphorus (TP)[a]	0.09	0.010
Cd[d]	1.11	0.151
Cu[d]	4.93	0.199
Fe[d]	161	3.377
Mn[d]	10.1	0.126
Ni[d]	19.2	0.203
Pb[d]	2.51	0.140
Zn[d]	3.07	0.083

[a] Units = mg L^{-1}
[b] Units = µS cm^{-1}
[c] Units = °C
[d] Units = µg L^{-1}

TABLE 6.11
Principal Components Analysis Report for the Physicochemical Lake Water Parameters[a]

Vector	Eigenvalue	Total Variance (%)	Cumulative (%)
1	4.2737	30.52	30.52
2	2.3285	16.63	47.15
3	1.7289	12.34	59.50
4	1.5604	11.14	70.65
5	1.1565	8.261	78.91
6	0.8222	5.873	84.78
7	0.7824	5.588	90.37
8	0.5177	3.698	94.07
9	0.4598	3.284	97.35
10	0.1576	1.126	98.48
11	0.1277	0.912	99.39
12	0.0668	0.477	99.87
13	0.0175	0.125	99.99
14	0.0004	0.003	100.0

[a] Only the first four eigenvalues were retained.

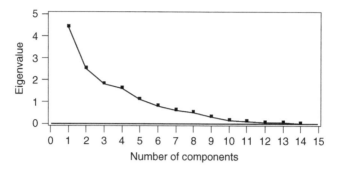

FIGURE 6.9 Scree plot for analysis of lake water parameters. Shown are the sorted eigenvalues as a function of the eigenvalues index.

flatten out, the corresponding components may be regarded as insignificant. These four were retained and used for Varimax rotation. Rotations are used to better align the directions of the factors with the original variables so that the factors may be more interpretable. Varimax, in particular, searches for an orthogonal rotation (i.e., a linear combination) of the original factors such that the variance of the loadings is maximized and the factors remain uncorrelated. This is opposite of oblique rotation where the resulting factors will be correlated. The Varimax rotated factor pattern for four PC extracted factors is shown in Table 6.12.

As shown in Table 6.12, PC 1 had higher loadings for conductivity, TDS, Cd, Cu, and Pb and may represent metal accumulation with particulate matter. PC 2 is

TABLE 6.12

Varimax Rotated Factor Pattern for Five PC Extracted Factors for the Physicochemical Lake Water Parameters

Parameter	Factor 1	Factor 2	Factor 3	Factor 4
pH	−0.245990	−0.736481	0.294309	0.168500
Dissolved oxygen (DO)	−0.408170	0.395762	−0.240894	0.596796
Conductivity	0.869445	0.054768	−0.024219	0.039072
Temperature	0.417058	−0.496904	0.337554	−0.628627
Total dissolved solids (TDS)	0.867444	−0.029485	0.311401	0.004895
Total nitrogen (TN)	0.093741	0.159563	0.669584	0.264504
Total phosphorus (TP)	−0.608351	0.043344	0.465586	0.139689
Cd	0.628564	0.366538	−0.164084	0.025129
Cu	0.665056	0.398710	−0.013808	0.156666
Fe	0.232938	−0.149541	0.661706	0.258328
Mn	0.130755	0.196907	0.472253	−0.531516
Ni	−0.290059	0.513149	0.702319	0.042803
Pb	0.919216	0.040824	−0.034966	0.282521
Zn	0.384059	−0.645934	0.003390	0.518140

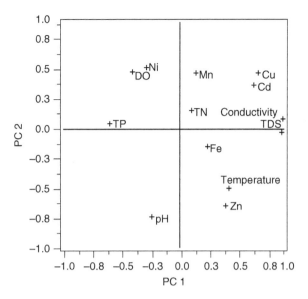

FIGURE 6.10 Factor loading plot for the first two PCs from PCA analysis of lake water parameters.

associated with Ni. PC 3 represents the carrier of micronutrients as higher loadings for TN, TP, Fe, and Mn were shown. PC 4 is associated with DO and Zn. The corresponding factor loading plot for the first two PCs is shown in Figure 6.10. The loadings plot shows a matrix of two-dimensional representations of factor loadings. From this, information regarding the correlation between variables can be obtained.

The representative score plot is shown in Figure 6.11. Score plots show a two-dimensional matrix representation of the scores for each pair of PCs. Such plots represent the linear projection of objects denoting the main part of the total variance of the data. As shown, there are a number of clusters representing the 15 lake water samples. Figure 6.12 displays a 3D spinning plot of the first two PCs. The variables show as rays (biplot rays) in the plot and approximate the variables as a function of the PCs on the axes. Note that the length of the rays correspond to the eigenvalue or variance of the PCs.

In summary, PCA rendered an important data reduction into four principal components describing 71% of the cumulative variance for lake water physicochemical parameters. Analysis revealed that such data consisted of three major components representing a carrier of micronutrients (TN, TP, Fe, and Mn), metal accumulation with dissolved particulate matter and one associated with DO and Zn. A fourth represented Ni alone.

6.5 PATTERN RECOGNITION—SUPERVISED

In supervised methods, the groupings of samples must be known to allow accurate predictions to be performed. More detailed explanations of these methods along with their applications are presented in detail in the following sections.

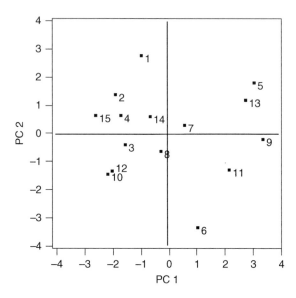

FIGURE 6.11 Score plot for PCA analysis of lake water parameters.

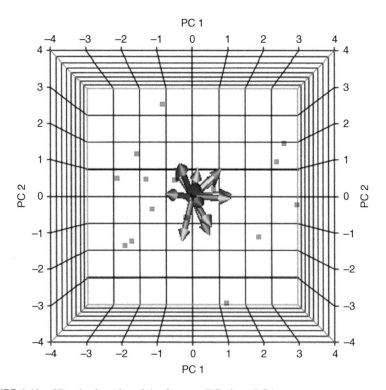

FIGURE 6.12 3D spinning plot of the first two PCs from PCA.

6.5.1 DISCRIMINANT ANALYSIS

Discriminant analysis (DA) is a supervised classification technique that is used in investigating and classifying a set of observations into predefined classes. Ostensibly, its goal is to classify each observation into one of the groups. Unlike PCA, it employs *a priori* knowledge in order to reach a solution. Here, the significance is based in terms of the means of the discriminating variables and operates on the assumption of raw data. Ostensibly, it is a special case of canonical correlation used to analyze dependence. As with factor analytic techniques described above, key concept definitions are needed to provide a basis for learning:

1. *Discriminating variables*—independent variables often termed predictors;
2. *Criterion variable*—the dependent variable;
3. *Discriminant function*—a latent variable that is created as a linear combination of discriminating (independent) variables and calculated for each group as (Wunderlin et al. 2001)

$$f(G_i) = k_i + \sum_{j=1}^{n} w_{ij} p_{ij} \qquad (6.16)$$

where i is the number of groups (G), k_i the constant inherent to each group, n the number of parameters used to classify the dataset into a given group, and w_j is the coefficient assigned by DA analysis to a given selected parameter (p_j).

4. *Discriminant coefficients*—used in the formula for making the classifications in DA, much as b coefficients are used in regression to make predictions. These come in the form of standardized and unstandardized coefficients and are used to classify importance of the independent variables.
5. *Discriminant score*— value resulting from applying a discriminant function formula to the data for a given test case. The standardized discriminant function score for the ith function is calculated as

$$D_i = d_{i1} z_1 + d_{i2} z_2 + \cdots + d_{ip} z_p \qquad (6.17)$$

where d_i is the discriminant function coefficient and z is the score on each predictor.

6. *Centroids*—group means on D_i.
7. *Tests of significance*—the Wilks' lambda test is typically used to test the significance of the discriminant function as a whole. Lambda varies from 0 to 1, with 0 meaning group means differ, and 1 meaning all groups are the same. Mahalanobis distances are used in analyzing cases in DA.
8. *Hold out sample*—often used to validate the discriminant function.

> The term *canonical analysis* is often confused due to multiple meanings in the literature. It is most commonly used as an extension of multivariate regression and is concerned with determining relationships between groups and variables (*X* and *Y*) in a given dataset. Note that this term is also synonymous with conical correlation and conical correspondence analysis.

Discriminant analysis is performed in a two-step process: testing significance of a set of discriminant functions and classification. Remember that DA is group specific, that is, many variables are employed in the study in order to determine the ones that discriminate among groups. The first step is computationally similar to multivariate analysis of variance (MANOVA) where matrices of total variance and covariances and pooled-within group variances and covariances are generated and compared via multivariate *F*-tests to determine significant group differences (Liu et al. 2008). Crossproduct matrices are first constructed for between-group (bg) differences and within-group (wg) differences:

$$SS_{total} = SS_{bg} + SS_{wg} \qquad (6.18)$$

The determinants for these matrices are then calculated and used to compute a test statistic, usually in the form of the Wilks' lambda:

$$\Lambda = \left| \frac{S_{wg}}{S_{bg} + S_{wg}} \right| \qquad (6.19)$$

Finally, an *F*-ratio is calculated as in MANOVA:

$$F_{t(k-1);ms-v} = \left[\frac{1 - \Lambda^{1/s}}{\Lambda^{1/s}} \right]\left[\frac{ms - v}{t(k-1)} \right] \qquad (6.20)$$

where t is the number of variables, k the number of treatments, $m = (2kn - t - 2)/2$, $s = [t^2(k-1)^2 - 4]/[t^2 + (k-1)^2 - 5]^{1/2}$, and $v = [t(k-1) - 2]/2$. Note that the degrees of freedom for *F* are $t(k-1)$ for the numerator and $ms - v$ for the denominator. The MANOVA can be reconceptualized as a DA and a significant *F* test can result in the determination of a discriminant function.

> *MANOVA* constructs a linear combination of dependent variables (DVs) and then tests for differences in the new variable using ANOVA type methods. The independent variable used to group the cases is categorical. Next, MANOVA tests whether the categorical explains a significant amount of variability in the new DV.

If the multivariate statistical test is significant, investigators can proceed to determine the variables that have significantly different means across the groups. If the group means are significant, classification of variables can proceed. Discriminant analysis allows for the determination of optimal combinations of variables with functions providing discrimination among groups. Classification is then possible from the canonical functions produced from canonical correlation analysis. Here, variables are classified into groups in which they have the highest classification scores.

There are two general rules applied to DA classification procedures with the simplest occurring if there are only two groups to classify. Here, one can classify based on the discriminant function scores—groups above and below 0 are grouped separately. When there are more than two groups, the following formula is used:

$$C_j = c_{j0} + c_{j1}x_1 + c_{j2}x_2 + \cdots + c_{jp}x_p \tag{6.21}$$

where the classification scores for group j is found by multiplying the raw score on each predictor (x) by its associated classification function coefficient (c_j), summing over all predictors and adding a constant, c_{j0}. Note that c_j is found by taking the inverse of the within-subjects covariance matrix W and multiplying by the predictor means:

$$c_j = W^{-1} M_j \tag{6.22}$$

Finally, the intercept is found by the following equation:

$$c_{j0} = -\frac{1}{2}C_j M_j \tag{6.23}$$

From an environmental perspective, a discriminant function may be interpreted by examining the correlations between the original (predictor) variables and the discriminant function. Note that the method used to enter the predictor variables is important with standard, forward stepwise, and backward stepwise techniques utilized. The standard mode constructs discriminant functions containing all predictor variables. In forward stepwise mode, the predictor variables are entered step-by-step in order of significance until no further gain in discrimination is achieved by the addition of more variables to the discriminant function. Finally, the backward stepwise mode removes predictor variables, step-by-step, beginning with the least significant, until no major changes are noticed. Ultimately, the use of beta coefficients and the structure matrix accounts for full interpretation, with the larger the coefficient, the greater the contribution of the given variable to the discrimination among groups. Examination of the factor structure allows investigators to identify the independent variables that contribute to the discrimination among dependent variables.

6.6   *K*-NEAREST NEIGHBOR CLASSIFICATION

K-nearest neighbor classification is a nonparametric discriminant method that utilizes multivariate distances to categorize an unknown sample of the validation

---

EXAMPLE PROBLEM 6.2

1. Although similar in nature, cluster analysis and discriminant analysis have key differences. Explain these differences in detail.

Answer: These techniques differ significantly in their training or learning period. In cluster analysis, an unsupervised learning technique, the groups (clusters) are not predetermined. The purpose of training is to know the category of each object. Discriminant analysis is a supervised learning technique where the groups are determined beforehand and the object is to determine the linear combination of independent variables that best discriminates among the groups. The main purpose of training is to know the classification rule.

---

set based on its proximity to those categorized on the training set. Euclidian distance ($d_{i,j}$) between two samples i and j is calculated as (Roggo et al. 2003)

$$d_{i,j} = \sqrt{\sum_{k=1}^{i=n} (x_{ik} - x_{jk})^2} \tag{6.24}$$

where n is the number of variables and x_{ik} is the value of the variable for the sample i. The k-closest neighbors (k = the number of neighbors) are then rounded by examining the distance matrix, with the k-closest data points analyzed to determine the most common class label among the set. Note that the k-neighbors of an unknown sample are the training samples that have the lowest Euclidian value. The prediction class is the one having the largest number of objects among the k-neighbors (Roggo et al. 2003).

How do investigators determine the optimum k number? This value is determined by a cross-validation procedure (Alsberg et al. 1997) where each object in the training set is removed and considered as a validation sample. The KNN test is run on $K - 1$, for example, where two or more class labels occur an equal number of times for a given data point within the dataset. If there is a tie between classes, KNN is run on $K - 2$ until $K = 1$. The resulting class labels are used to classify data points within the dataset. Note that data preprocessing is commonly required to avoid the effect of differing scales of the studied variables.

As with our discussion on LDA, KNN does not work efficiently if large differences in the number of samples are present in each class (Berrueta et al. 2007). In addition, KNN provides relatively poor information about the structure of the classes and does not provide a graphical representation of the results. Despite these limitations, KNN has been successfully utilized in a variety of environmental applications. For example, Martín et al. (2008) employed KNN to predict carbon monoxide ground level concentrations in the Bay of Algeciras, Spain. More specifically, the KNN classification technique was used to identify peaks of carbon monoxide concentrations for subsequent analysis by neural network methods (see Section 6.11.1).

6.7   SOFT INDEPENDENT MODELING OF CLASS ANALOGY

Soft independent modeling of class analogy (SIMCA) is one of the most widely used class-modeling techniques employed in environmental analysis. Unlike the KNN method, SIMCA classification employs a PCA model based on selected observations and determines class membership by calculating the distance between a new observation and the PCA model (Wold et al. 1986). The number of principal components for each class in the training set is determined by cross-validation to account for the majority of the variation within each class (Berrueta et al. 2007). In SIMCA, the class distance is traditionally calculated as the geometric distance from the principal component model. Hence, SIMCA results can be graphically displayed, a major advantage over the KNN method.

Lundstedt-Enkel et al. (2006) used SIMCA classification to investigate the concentrations of organochlorines and brominated flame-retardants (BFRs) in Baltic Sea guillemot (*Uria aalge*) egg and muscle tissues. This work aimed to assess the relationship between egg and adult birds regarding concentrations of the chosen contaminants. SIMCA was used to investigate if the contaminant pattern in the two matrices differed significantly from each other. Results show that the two matrices were completely separated into two classes and were thus totally different.

6.8   MULTIVARIATE CALIBRATION METHODS

The techniques of linear regression, discussed in Chapters 2 and 4, can be extended to form multiple regression models with the general form

$$y = \beta_0 + \beta_1 X_1 + \beta_2 X_2 + \cdots + \beta_k X_k + \varepsilon \qquad (6.25)$$

Here, one response variable, *y*, depends on a number of prediction variables (X_1, X_2, X_3, ... etc.). As we learned with univariate regression, residual analysis is important for model evaluation purposes. There is also the option of omitting chosen predictor variables with the ultimate ability to try all possible combinations to identify the one that predicts *y* successfully (Miller and Miller 2000).

The development of multivariate calibration models has been instrumental in recent years to explain global spectral variance and for the analysis of unresolved peaks in chromatographic separations (Tan and Brown 2003, Azzouz and Tauler 2008). Table 6.13 lists modern environmental applications utilizing a variety of multivariate calibration techniques. A more detailed examination of the commonly utilized techniques is presented in the following sections.

6.8.1   MULTIVARIATE LINEAR REGRESSION

Multivariate linear regression (MLR) is likely the simplest computational multivariate calibration model and normally applied when no explicit causality

TABLE 6.13

Selected Environmental Applications Employing Various Multivariate Calibration Models

Model	Model Usage/Approach	Defining Characteristics/Results	Reference
MLR	MLR used to resolve overlapping of absorbance spectra recorded during elution of analysis.	Online simultaneous spectrometric determination of nitro-substituted solid-phase extraction. Successful extraction and determination of 2-, 3-, and 4-nitrophenol.	Manera et al. (2007)
MLR	MLR used to build models to predict relationships between stomatal conductance and environmental factors.	Study to address important aspects of plant–environment interactions such as water relations and pollutant uptake.	Vitale et al. (2007)
MLR	Use of MLR models to predict ambient air concentration of polycyclic aromatic hydrocarbons (PAHs).	MLR models aided to relate outdoor ambient $PM_{2.5}$ and PAH concentrations for epidemiological studies based on publicly available meteorological and $PM_{2.5}$ data.	Lobscheid et al. (2007)
PLS	PLS-1 developed using infrared (IR) spectra of unresolved chromatographic peaks.	The PLS-1 model was successfully used to predict concentrations of 2,4- and 2,5-dichlorophenol in water samples.	Rodriguez et al. (1996)
PLS	Development of partial least squares (PLS) models for the prediction of oxygen uptake and humic acid content in composts based on Fourier transform infrared (FTIR) spectra and reference values.	Validated FTIR-PLS prediction models for respiration activity and for humic acid content allows appropriate evaluation to assess compost stability. This presents an effective analytical method for waste management purposes.	Meissl et al. (2007)
PLS	Use of PLS models to predict the effects of experimental design, humidity, and temperature on polymeric chemiresistor arrays.	PLS models were shown to quantitatively predict trichloroethylene (TCE) levels using an array of chemiresistors. The effects of experimental design, humidity, and temperature were considered.	Rivera et al. (2003)
PLS and PCR	Use of PLS and principal component regression (PCR) kinetic models for the simultaneous determination of Fe^{2+} and Fe^{3+}.	PLS and PCR models successfully applied for the simultaneous determination of Fe^{2+} and Fe^{3+} in environmental samples. This approach allowed the transformation of the two iron oxidation states to be monitored over time.	Absalan and Nekoeinia (2005)
PCR	PCR method used to spectrophotometrically determine the concentration of eluting dyes in dyehouse wastes.	PCR helped determine trace sulphonated and azosulphonated dyes in dyehouse wastes. Accomplished without prior separation from the waste mixture, ultimately negating the need for sample preparation tasks.	Al-Degs et al. (2008)

between dependent and independent variables is known. This technique, however, does suffer from a number of limitations including overfitting of data, its dimensionality, poor predictions, and the inability to work on ill-conditioned data (Walmsley 1997). Note that if the number of descriptive variables exceeds the number of objects, collinearity is likely to occur. In this situation, investigators can make use of variable selection prior to modeling with MLR.

Two methods are commonly employed to aid variable selection: stepwise multiple linear regression (stepwise MLR) and genetic algorithm multiple linear regression (GA-MLR). Stepwise MLR is used to select a small subset of variables from the original matrix $\mathbf{X}$. It starts with a given variable that has the highest correlation with the response variable (e.g., CO_2 concentration level). If the given variable results in a significant regression (evaluated with an overall F-test), the variable is retained and selection continues (Deconinck et al. 2007).

Next, the variable that produces the largest significant increase of the regression sum of squares (evaluated with a partial F-test) is added. This is termed the forward selection procedure. After each forward selection step, the significance of the regression terms already in the model are tested, with the nonsignificant terms eliminated. This is termed the backward elimination step.

In the GA-MLR method, a population of k strings is randomly chosen from the predictor matrix $\mathbf{X}$. Each string consists of a row-vector with elements equal to the number of descriptive variables in the original dataset. The fitness of each string is evaluated with the use of the root mean squared error of prediction formula (Estienne et al. 2000):

$$\text{RMSEP} = \sqrt{\sum_{i=1}^{n_t} \frac{(\bar{y}_i - y_i)^2}{n_t}} \tag{6.26}$$

where n_t is the number of objects in the test set, y_i the known value of the property of interest for object i, and $\bar{y}$ is the value of the property of interest predicted by the model for object i. In simplest terms, the strings with the highest fitness are selected and repeatedly subjected to crossovers and mutations until investigators are content with the model statistics.

6.8.2 PARTIAL LEAST SQUARES

Partial least squares (PLS) is one of the soft-model-based regression techniques that utilize regressing eigenvectors from the data matrix onto the dependent variable. A related technique, principal component regression (PCR), is covered in Section 6.8.3. Recall our discussion of the PCA method in Section 6.4.2. PLS is an extension of PCA and utilizes an iterative algorithm that extracts linear combinations of essential features of the original $\mathbf{X}$ data while modeling the $\mathbf{Y}$ data dependence on the dataset (Amaral and Ferreira 2005).

The PLS method uses latent variables u_i (matrix $\mathbf{U}$) to model objects separately on the matrix of $\mathbf{Y}$-dependent data and t_i variables (matrix $\mathbf{T}$) for modeling objects separately in the **X** matrix of independent data. The latent variable **U** and **T** are used to form the regression model determined by

$$U = A \times T + E \tag{6.27}$$

in an iterative process with the centered matrices of **X** and **Y** as starting points (Amaral and Ferreira 2005).

In terms of errors, PLS models calculate goodness of fit based on the error of the prediction. Cross-validation tests are used to determine the number of significant vectors in U and T. Additionally, they are used to determine the overall error of prediction. In the cross-validation routine, one sample is left out of the calibration set followed by the construction of a calibration model. Finally, the concentration of the sample left out is predicted. Note that data preprocessing techniques (e.g., mean centering) can be used to improve precision in this process.

A closer look at the study by Rivera et al. (2003) highlighted in Table 6.13 can assist PLS interpretation. This study utilized PLS to provide quantitative predictions of trichloroethylene (TCE) using an array of chemiresistors through appropriate experimental design. Chemiresistors are a class of chemical sensors that has shown great promise for VOC determination in various environmental matrices (Ho and Hughes 2002). The objective of this study was to assess the effects of temperature and humidity on the calibration of the chemiresistors for designed (D) and non-designed (ND) calibrations. As discussed in Chapter 4, properly designed experiments are paramount. They can also greatly enhance the predictive ability of multivariate calibrations (Thomas and Ge 2000).

Table 6.14 lists the calibration results for temperature and humidity for both designed and nondesigned experiments. Note that E1–E4 represent: E1 = 31°C and 0% relative humidity; E2 = 31°C and 100% relative humidity; E3 = 23°C and 0% relative humidity; E4 = 23°C and 100% relative humidity. Also note that cross-validated standard error of prediction (CVSEP) is the cross-validated root mean square error between the known concentration and the predicted concentration for a

---

TABLE 6.14

PLS Calibration Data for Designed and Nondesigned Sequences

Environmental Condition	r^2	CVSEP (ppm)	Number of Factors
31°C, 0/100% humidity (E1 + E2)	0.980 (D)	400	2
	0.995 (ND)	230	2
23°C, 0/100% humidity (E3 + E4)	0.980 (D)	450	2
	0.970 (ND)	480	2
23/31°C, 100% humidity (E2 + E4)	0.970 (D)	590	3
	0.980 (ND)	430	4
23/31°C, 0% humidity (E1 + E3)	0.940 (D)	750	2
	0.910 (ND)	980	2

Source: Modified from Rivera, D., et al., *Sensors and Actuators B*, 92, 110–120, 2003. With permission from Elsevier.

TABLE 6.15

PLS Prediction Data for Designed and Nondesigned Sequences

Environmental Condition	r^2	SEP (ppm)
31°C, 0/100% humidity (E1 + E2)	0.96 (D)	730
	0.71 (ND)	840
23°C, 0/100% humidity (E3 + E4)	0.96 (D)	730
	0.90 (ND)	1000
23/31°C, 100% humidity (E2 + E4)	0.97 (D)	590
	0.92 (ND)	890
23/31°C, 0% humidity (E1 + E3)	0.85 (D)	1200
	0.74 (ND)	1600

Source: Modified from Rivera, D., et al., *Sensors and Actuators B*, 92, 110–120, 2003. With permission from Elsevier.

given cross-validated calibration model. These models were used to predict data collected under the same environmental conditions for the alternate design. Note that standard error of prediction (SEP) is the root mean square error between the reference concentration and the predicted values. The investigators found that for the stated environmental conditions, nondesigned calibration models had poorer prediction statistics (Table 6.15) than that expected from the calibration statistics.

Figure 6.13 presents the prediction results using the model from the nondesigned method to predict the designed data (Figure 6.13a). Figure 6.13b shows the prediction using designed experiments to predict the nondesigned data. Note that the 23/31°C, 100% humidity scenario was used for both. The investigators stressed that prediction with nondesigned data showed a greater spread of data in the 1000–5000 ppm range than prediction with designed data.

The investigators also commented that the standard deviation of the predicted data in the 1000–5000 ppm range was 1.9 times greater for the set predicted with the model from the nondesigned data compared with the set predicted from designed data. Is there a logical explanation for this? Does this differ from what is discussed in Chapter 4 in terms of the importance of properly designed experiments? The investigators believe that the cause of the reduced variance using designed data at lower concentrations is likely due to the fact that those models have greater independence of hysteresis and baseline drift than nondesigned models.

6.8.3 PRINCIPAL COMPONENT REGRESSION

As discussed in Section 6.4.2, scores on a number of principal components can be used as descriptive variables and help find structure within datasets. Such characteristics can be used in MLR in a process termed *principal component regression* (PCR). Rather than forming a single model as with MLR, investigators have the choice of forming models using 1, 2, or more components. As discussed, this has obvious advantages over MLR as underlined structure can be determined. Principal components are selected in decreasing order of variance and are uncorrelated in

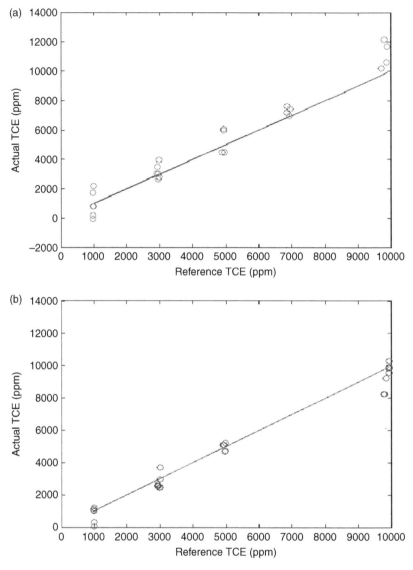

FIGURE 6.13 Trichloroethylene (TCE) prediction results using: (a) nondesigned PLS models and (b) designed PLS models to predict the nondesigned data. (From Rivera, D., et al., *Sensors and Actuators B*, 92, 110–120, 2003. With permission from Elsevier.)

nature. A major problem with MLR is that the variables are correlated, ultimately causing instability. In PCR, there is the ability to remove the original variables containing collinearity and thus help guarantee stable models.

So how does PCR progress? The original data matrix **X** is approximated by a small set of orthogonal PCs. Decisions are then made as to how many PCs to retain. This is often done according to the percentages of variance that each PC explains.

TABLE 6.16

Total Variance Explained for the Five Components for the Study of Environmental Factors on Esophageal Cancer Mortality

Component	Eigenvalues	Variance (%)	Cumulative Variance (%)
1	3.956	30.431	30.431
2	2.154	16.568	46.499
3	1.974	15.187	62.186
4	1.567	12.052	74.238
5	1.427	10.993	85.231

Source: From Wu, K.-S., Huo, X., and Zhu, G.-H., *Science of the Total Environment*, 393, 219–225, 2008. With permission from Elsevier.

This process is termed PCR top-down. It must be noted that the PCs explaining the largest variance in **X** are not always related to **Y** (Estienne et al. 2000). A leave one out (LOO) cross-validation method is also utilized. Here, predictive ability of the model is estimated (1, 2, or more PCs) in terms of root mean squares errors of the cross-validation (RMSECV). Detailed procedures behind this approach can be found in a variety of informative studies (Wold 1978, Verdú-Andrés et al. 1999, Estienne et al. 2000). Regardless of the approach, an MLR model is then built that relates the scores of the PCs to the response or property of interest.

Principal component regression [in combination with the Geographical Information System (GIS)] was instrumental in studying the relationship between esophageal cancer and 13 spatial environmental factors in China (Wu et al. 2008). The investigators hypothesized that temporal and spatial variation in climate and geographical factors were associated with esophageal cancer (mortality). To test this hypothesis, they performed PCA to combine information on spatial environment followed by equamax rotation of the factor loading solution. Components were extracted and MLR performed using the selected components as independent variables and esophageal cancer mortality as the dependent variable.

As shown in Table 6.16, the original 13 variables were reduced to five new factors via PCA. These new components were described in terms of the original variable loads (Table 6.17). Take component 4, for example. Inspection reveals that this component represented DI, WHI, and NDVI of July. Component 5 represented altitude and component 1 precipitation, mean temperature, lowest temperature, highest temperature, and wind speed. From this data, investigators concluded that the original variables: precipitation, temperature, wind speed, elevation, DI, WHI, and NDVI all had significant associations with esophageal cancer mortality. Table 6.18 presents the results of components 1, 4, and 5 entered into the MLR equation for both male and female subjects.

6.9 MULTIVARIATE T^2 CONTROL CHARTS

The traditional control charts discussed in Chapter 3 cannot be used for many environmental monitoring applications because they do not handle multivariate data,

TABLE 6.17

Equamax Rotated Component Matrix for the Study of Environmental Factors on Esophageal Cancer Mortality

Variable	Component 1	Component 2	Component 3	Component 4	Component 5
Precipitation (mm)	0.833	—	0.320	0.292	—
Mean temp (°C)	0.907	0.144	0.337	—	0.131
Mean lowest temp (°C)	0.910	0.124	0.311	—	0.143
Mean highest temp (°C)	0.899	—	0.387	−0.160	—
Wind speed (m/s)	−0.531	0.184	0.520	—	0.438
DI	−0.566	—	—	−0.635	—
WHI (mm/°C)	−0.265	—	—	0.798	−0.185
NDVI of July	—	−0.523	—	0.548	0.426
NDVI of Jan	0.282	−0.160	0.745	0.170	−0.317
Evaporation (mm)	0.229	—	0.822	−0.257	0.203
Altitude (m)	−0.116	−0.171	—	—	−0.889
GDP per km^2	—	0.938	—	—	0.123
Population per km^2	—	0.931	—	—	0.165

DI = Drought Index, WHI = Water-Heat Index, NDVI = Normalized Difference Vegetation Index, GDP = Gross Domestic Product.

Source: From Wu, K.-S., Huo, X., and Zhu, G.-H., *Science of the Total Environment*, 393, 219–225, 2008. With permission from Elsevier.

TABLE 6.18

MLR Analysis between Logarithm of Esophageal Cancer Mortality and Extracted Compounds

Components	B[a]	Beta[b]	r^2	F	p
Male					
PC4	−0.121	−.0245	0.117	10.240	<20.001
PC5	0.092	0.186			
PC1	−0.074	−0.150			
Female					
PC4	−0.101	−0.190	0.061	7.524	<20.001
PC5	0.84	0.159			

[a] B = unstandardized coefficients,
[b] Beta = Standardized coefficients

Source: From Wu, K.-S., Huo, X., and Zhu, G.-H., *Science of the Total Environment*, 393, 219–225, 2008. With permission from Elsevier.

especially with regard to correlated factors. In such instances, multivariate control charts are useful tools for monitoring, for example, where and when an impact may have occurred or, once detected, may still be occurring. A common method to maintain control and to monitor such events is the construction of a multivariate control chart based on Hotellings's T^2 statistic defined as

$$T^2 = (Y - \mu)'S^{-1}(Y - \mu) \qquad (6.28)$$

$$T^2 = n(\overline{Y} - \mu)'S^{-1}(\overline{Y} - \mu) \qquad (6.29)$$

for ungrouped and grouped data, respectively, where S is the covariance matrix, μ the true mean, Y and $\overline{Y}$ the observations, and n is sample size. The T^2 chart is plotted along with control limits that are determined combining relevant values, either from past data or for the desired target values with critical values from a normal distribution.

There are two distinct phases when constructing a multivariate T^2 chart. Phase 1 involves testing whether a process is in control based on the initial individual or subgroup data (no target value specified) during a specific time interval. This phase establishes control limits for purposes of monitoring with the upper control limit (UCL) a function of the β distribution. Phase two utilizes the target value to monitor the behavior of the process by detecting any departure from the target as future samples are drawn. Here, the UCL is a function of the F-distribution.

As mentioned, multivariate T^2 charts are useful for data where correlation between multiple factors is present and can be used to detect events (e.g., storm events, nonpoint source releases) that may result in elevated levels of one or more parameters in related environmental systems. Consider the hypothetical dataset in Table 6.19, which presents the physicochemical parameter results of an initial study (phase 1) of the analysis of 20 river water samples over a given time frame. A T^2 chart constructed for phase 1 is presented in Figure 6.14a. As shown, the process seems to be in good statistical control since there is only one out-of-control point. Therefore, this chart can be used to establish targets for subsequent monitoring of this aquatic system. The phase 2 T^2 chart (Figure 6.14b) highlights monitoring of the subsequent process based on the generation of test statistics (Table 6.20) from phase 1.

A problem with the above approach lies in the fact that when the T^2 statistic exceeds the UCL, investigators cannot fully discern the particular parameter(s) that cause out-of-control results. A way to help alleviate this is to construct charts using the relative principal components obtained from PCA of a given dataset. As discussed, PCs are used to reduce dimensionality since the first two or three PCs typically account for the majority of the variability. Figures 6.15a and b present phase 1 and phase 2 T^2 charts constructed from the relative PCs generated from the hypothetical aquatic monitoring study presented earlier. The first two PCs accounted for 84% of the variation. The analysis of the projections (scores, Figure 6.16a) of the original data (observations at different times) and composition of each PC in terms of the contribution of the original factors (loadings, Figure 6.16b) can aid the identification of the causes that produced the out of control points noticed.

TABLE 6.19

Physicochemical Parameters of the Analysis of 20 River Water Samples Used in Multivariate T^2 Control Chart Analysis

Sample	River Flow $(m^3 s^{-1})$	Nitrate (μM)	TP (μM)	pH
1	7.14	4.67	3.44	7.55
2	7.14	4.44	3.56	7.55
3	7.15	3.97	3.67	7.60
4	7.12	4.42	3.34	7.70
5	7.20	4.34	3.99	7.56
6	7.34	4.34	3.87	7.65
7	7.67	4.55	3.77	7.34
8	8.02	5.41	3.88	7.27
9	7.59	5.03	4.23	7.15
10	8.00	6.40	4.35	7.21
11	7.34	5.03	4.30	7.11
12	7.54	5.01	4.03	7.09
13	7.89	4.76	3.99	7.12
14	8.00	4.78	3.97	7.17
15	7.67	4.56	4.21	7.20
16	7.76	4.36	3.78	7.32
17	7.56	4.55	3.78	7.31
18	7.97	3.99	3.67	7.28
19	7.54	3.99	3.77	7.29
20	7.21	4.01	3.77	7.36

6.10   CUSUM CONTROL CHARTS

Multivariate T^2 control charts are arguably the most widely used tools for monitoring multiparameter data for statistical control, but do have their limitations. Such charts are not entirely sensitive to small changes (one σ shift) of the process mean and the scale of the values displayed on the chart are not directly related to the scales of any of the monitored variables (Robotti et al. 2007). For these reasons, the Cumulative Sum (CUSUM) chart is a reliable alternative, and valuable when used in combination with Shewhart charts.

   This chart utilizes the cumulative deviation of each successive measurement from a target value (e.g., mean) and related statistical parameters (e.g., k, α) as discussed in Chapter 3. CUSUM charts can be one-sided in that they detect a shift in one direction from the target mean, or two-sided in that they detect a shift in either direction. Consider a hypothetical study of 50 sampling sites in a given watershed, with an environmental factor measured at all the sites over a period of time. The goal is to evaluate whether there is variability in the environmental factor over time at the various sites. Further consideration can then be given to the evaluation of spatial correlation among the sites.

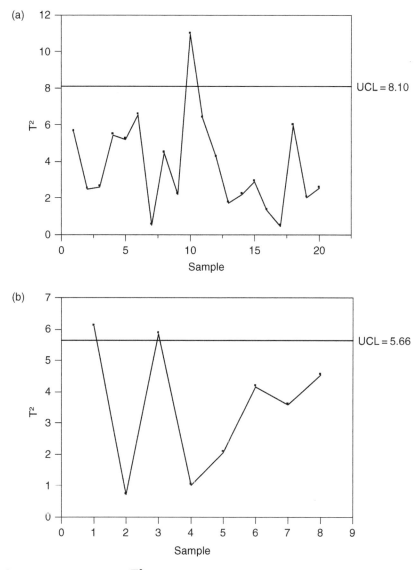

FIGURE 6.14 Multivariate T^2 charts for water quality parameters: (a) phase 1 and (b) phase 2 based on test statistics generated from phase 1.

Figure 6.17 presents the pattern obtained when producing a two-sided CUSUM chart for this study. For interpretation, compare the data points with limits that compose a V-mask formed by plotting V-shaped limits. Note that many CUSUM charts employ horizontal decision intervals rather than the V-mask. The origin of the V-mask is the most recently plotted point on the chart with the arms extended backward on the x-axis. As shown, there are a few points that are out of control, with the majority of those crossing the lower arm. Otherwise, the process is in good

TABLE 6.20

Test Statistics Generated for the Phase 1 T^2 Multivariate Chart Analysis

Test Statistic	River Flow ($m^3 s^{-1}$)	Nitrate (μM)	TP (μM)	pH
Sample size	20	20	20	20
Mean	7.54	4.63	3.68	7.30
Standard deviation	0.32	0.57	0.27	0.19
Correlation—river flow	1.00	0.46	0.45	−0.69
Correlation—nitrate	0.47	1.00	0.60	−0.46
Correlation—TP	0.45	0.61	1.00	−0.71
Correlation—pH	−0.69	−0.45	−0.71	1.00

statistical control. Note that CUSUM charts are often used in parallel with T^2 control charts and can also utilize the advantages of PCA for construction.

6.11   SOFT COMPUTING TECHNIQUES

The development of advanced soft computing techniques are fast emerging given the current trends in the development of state-of-the-art analytical techniques and the vast amount of data generated as a result of the analysis of complex environmental systems. Soft computing differs from conventional (hard) computing in that, unlike hard computing, it is tolerant of imprecision, uncertainty, partial truth, and approximation (Saridakis and Dentsoras 2008). It can provide information for product and process design, monitoring, and control and for interpretation of very complex phenomena for which conventional methods have not yielded low cost, analytic, and complete solutions. The more commonly utilized methods of soft computing are discussed in greater detail below.

6.11.1   Artificial Neural Networks

Artificial neural networks (ANNs) is one of the most popular soft computing methods utilized in environmental analysis and is based on our understanding of the brain and the associated nervous system. A crude analogy can be visualized by the generalized ANN architecture presented in Figure 6.18, composed of three layers of interconnected neurons (nodes), each of which connects to all the nodes in the ensuing layer. There are input and output layers representing data inputs into the neural network and response (or often multiple responses) of the network to the inputs, respectively. One or more intermediate layers, termed hidden layers, may exist between the input and output layer, which are utilized to compute associations between parameters.

Artificial neural networks have been characterized more like robust nonlinear regression models used to expose underlying relationships in a dataset using pattern recognition theory (Huitao et al. 2002). In addition, ANNs are considered more

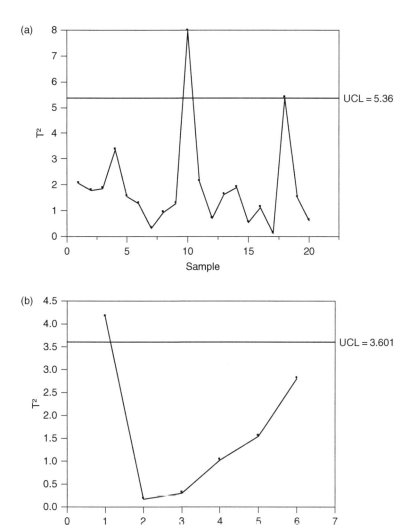

FIGURE 6.15   Multivariate T^2 charts constructed using relative PCs obtained for the water quality dataset: (a) phase 1 and (b) phase 2.

robust than other computational tools in solving problems in the following seven categories (Basheer and Hajmeer 2000):

1. Optimization (e.g., finding a solution that maximizes or minimizes an objective function).
2. Association (e.g., development of a pattern associator ANN to correct corrupted or missing data).
3. Control (e.g., designing a network that will aid in generating required control inputs).

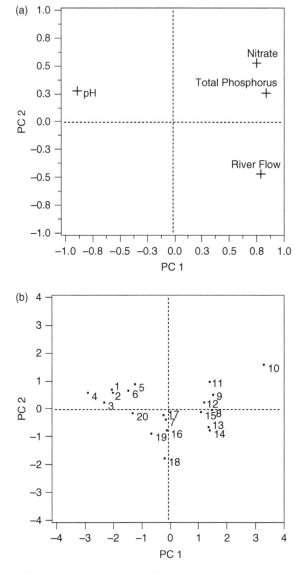

FIGURE 6.16   PCA analysis of the water quality dataset: (a) score plot and (b) loadings plot. Such plots can aid in the identification of the causes that produced out of control points.

4. Forecasting (e.g., training of an ANN for time series data to forecast subsequent behavior).
5. Function approximation (e.g., training ANN input–output data to approximate the underlying rules relating inputs to outputs).
6. Clustering (e.g., clusters are formed by exploring the similarities or dissimilarities between the input patterns based on their inter-correlations).
7. Pattern recognition (e.g., assigning an unknown input pattern to one of several prespecified classes).

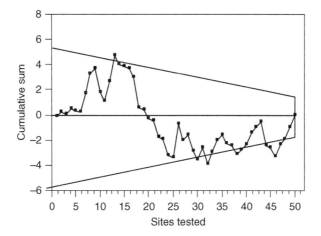

FIGURE 6.17 Two-sided CUSUM chart of a given environmental factor in the study of sampling sites in a watershed. The inserted V-mask represents the most recently plotted point on the chart with arms extended backward on the x-axis.

Further classification of ANNs are based on the degree of connectivity of the network neurons, the direction of flow of information within the network (recurrent or nonrecurrent), the type of learning algorithm, the learning rule, and the degree of learning supervision needed for training (Upadhaya and Eryureka 1992, Basheer and Hajmeer 2000). Think of an ANN as a collection of processing units that communicate by sending signals to each other over a number of weighted connections. Most units in ANN transform their net input by using a scalar-to-scalar function called an activation function. The activation value is fed via synaptic connections to one or more other units. An S-shaped activation function used to scale is the logistic function (JMP 2007):

$$S(x) = \frac{1}{1 + e^{-x}} \tag{6.30}$$

that scales values to have mean = 0 and standard deviation = 1. More detailed coverage of network calculations can be found in a number of valuable resources (Hopfield 1982, Widrow and Lehr 1990, Basheer and Hajmeer 2000, Samarasinghe 2006).

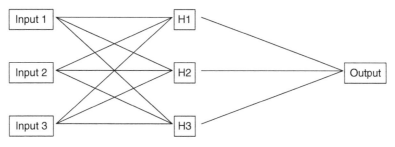

FIGURE 6.18 Generalized ANN architecture.

Note that ANNs are useless for the investigator unless they are well trained. Network learning is accomplished by two main kinds of algorithms: supervised and unsupervised. In supervised learning, desired outputs are known and incorporated into the network during training so the weights can be adjusted to match its outputs to the target values. After training, the network is tested by giving it only input values, not target values, and examining how close it comes to the correct target values. In unsupervised learning, the network lacks the correct output results during training. Unsupervised learning has undergone considerable study; nonetheless it is reported to lack broad application (Widrow and Lehr 1990). One exception is the use of a self organizing map (SOM), a type of ANN that is trained using unsupervised learning to produce low dimensional (one or two-dimensional) representations (maps) of the input space of the training samples (Kohonen 1997).

The most commonly reported algorithm that operates under supervised conditions is backpropagation, conceived by Rumelhart et al. (1986). This algorithm involves running each input pattern in a training set through a feedforward ANN, calculating the error incurred by the difference in actual and target network output, backpropagating (propagating errors in the same manner as forward signal propagation but in the opposite direction) and adjusting weights according to the error (Hopfield 1982). This process is repeated iteratively until the total error across the training set is below a specified maximum. See the following application for the use of additional algorithms in ANN models.

Investigators must take great care in choosing the number of nodes per layer and maximum training error because inappropriate selections will lead to over/underfitting of training set pattern data. Poor generalization is a symptom of an overfitted network, whereas poor classification of both training set and never-before-seen patterns occurs with an underfitted network (Basheer and Hajmeer 2000). Overfitting is especially detrimental to analyses as it can easily lead to predictions that are far beyond the range of the training data. Note that underfitting can also produce erroneous predictions. Effective approaches commonly used to avoid over/underfitting include model selection, jittering, weight decay (penalty coefficient), early stopping, and Bayesian estimation (Hopfield 1982).

Kassomenos and colleagues utilize an ANN model in their estimation of daily traffic emissions of CO, benzene, NO_x, PM_{10}, and VOCs in a South-European urban center (Kassomenos et al. 2006). As shown in Figure 6.19, the ANN model utilizes three layers: an input layer consisting of eight neurons (one for each input parameter), a hidden layer consisting of 20 neurons that implemented the S-shaped transfer function and an output layer consisting of five linear neurons that correspond to the five predicted emission rates. Network training for this model is evaluated using a variety of algorithms including Bayesian regularization (BR), resilient backpropagation (RB), scaled conjugate gradient (SCG), Broyden, Fletcher, Goldfarb, and Shanno (BFGS), and the Levenberg–Marquart (LM) algorithm.

The ANN model evaluation parameters for this study, root mean square error (RMSE) and relative root mean square error (RRMSE) are presented in Table 6.21. The mean percentage error and predicted values for each pollutant range are shown visually in Figure 6.20. The performances of the different ANN architectures and algorithms are shown in Table 6.22. From the results, it is evident that using Bayesian

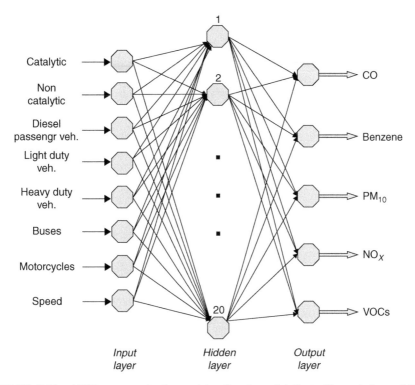

FIGURE 6.19 ANN structure in the model estimation of daily traffic emissions of CO, benzene, NO_x, PM_{10}, and VOCs in a South-European urban center. (From Kassomenos, P., Karakitsios, S., and Papaloukas, C., *Science of the Total Environment*, 370, 480–490, 2006. With permission from Elsevier.)

regularization constantly resulted in better results and the use of 10 to 20 hidden neurons allowed for successful modeling.

6.11.2 Fuzzy Logic

The concept of fuzzy logic is contrary to classical logical thought where the possibilities of a given conclusion are either true or false. In fuzzy logic, knowledge is represented

TABLE 6.21

ANN Evaluation Parameters for the Estimation of Daily Traffic Emissions

Test Statistic	CO	Benzene	PM_{10}	NO_x	VOCs
RMSE (g km^{-1} s^{-1})	0.03	0.004	0.0004	0.005	0.008
RRMSE (%)	0.006	0.002	0.0004	0.001	0.002

Source: Modified from Kassomenos, P., Karakitsios, S., and Papaloukas, C., *Science of the Total Environment*, 370, 480–490, 2006. With permission from Elsevier.

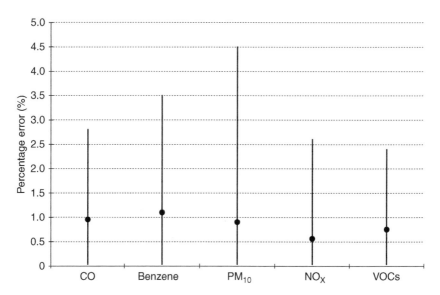

FIGURE 6.20   Mean percentage error and predicted values for traffic emission parameters. (From Kassomenos, P., Karakitsios, S., and Papaloukas, C., *Science of the Total Environment*, 370, 480–490, 2006. With permission from Elsevier.)

by if-then rules with differing grades of truth existing between true and false (Sproule et al. 2002). Fuzzy logic can be used for mapping inputs to corresponding outputs and are divided into four generalized parts (Ocampo-Duque et al. 2006):

1. *Fuzzification*—definition of inputs/outputs and membership functions.
2. *Weighting*—membership functions associate a weighting factor with values of each input that determine the degree of membership.
3. *Evaluation of inference rules*—applications of fuzzy operations to multi-part antecedents, the application of implication methods from the antecedent to the consequent for every rule, and the use of an aggregation method to join the consequents across all rules.
4. *Defuzzification*—the transformation of the fuzzy output into a nonfuzzy numerical value.

Full details of this process are beyond the scope of this book. For those interested in a more detailed look into fuzzy logic and associated computational structure, refer to the work of Pedrycz and Gomide (1998). Do consider, however, a generalized look into computational structure through the use of fuzzy sets with associated variables having a membership function (μ) between zero and one (Zadeh 1965). For an environmental variable, X, a fuzzy set A of acceptable values can be considered. If x is one possible value of X, then $A(x)$ denotes its membership degree in A (Adriaenssens et al. 2004). Ostensibly, a value of zero represents complete non-membership, the value one represents complete membership and values in between are used to represent intermediate degrees of membership.

TABLE 6.22

Performance of the Different ANN Architecture and Algorithms Used in the Evaluation of Traffic Emissions

Architecture	Training Algorithm	Performance (Mean Absolute Error) ($g\ km^{-1}\ s^{-1}$)
$8 \times 20 \times 5$	BR	0.0097
$8 \times 20 \times 5$	LM	0.0107
$8 \times 20 \times 5$	BFGS	0.0092
$8 \times 20 \times 5$	SCG	0.0096
$8 \times 20 \times 5$	RB	0.0191
$8 \times 10 \times 5$	BR	0.0089
$8 \times 10 \times 5$	LM	0.0098
$8 \times 10 \times 5$	BFGS	0.0092
$8 \times 10 \times 5$	SCG	0.0108
$8 \times 10 \times 5$	RB	0.0184
$8 \times 30 \times 5$	BR	0.0135
$8 \times 30 \times 5$	LM	0.0184
$8 \times 30 \times 5$	BFGS	0.0099
$8 \times 30 \times 5$	SCG	0.0093
$8 \times 30 \times 5$	RB	0.0198
$8 \times 5 \times 5$	BR	0.0133
$8 \times 5 \times 5$	LM	0.0113
$8 \times 5 \times 5$	BFGS	0.0134
$8 \times 5 \times 5$	SCG	0.0160
$8 \times 5 \times 5$	RB	0.0166

Source: Modified from Kassomenos, P., Karakitsios, S., and Papaloukas, C., *Science of the Total Environment*, 370, 480–490, 2006. With permission from Elsevier.

The concept of a fuzzy set can be schematically illustrated by a generalized example (Figure 6.21) using the water quality parameter, dissolved oxygen (DO). The x-axis represents the environmental parameter with the y-axis showing the membership function. Consider a DO value of 2.0 mg/L, which, as shown, belongs to both low and medium fuzzy sets, with membership functions of 0.2 and 0.4, respectively. It is thus possible for variables to belong to more than one set. Other water quality parameters (e.g., pH, temperature) can be integrated with DO through fuzzy reasoning to establish, for example, water quality indexes for environmental management and decision-making (Astel 2007, Ocampo-Duque et al. 2007).

So what are the advantages of fuzzy logic over other soft computing techniques? Studies have shown that fuzzy logic systems provide a more transparent representation of the system being investigated, making it a more flexible model and easily updated with new knowledge (Adriaenssens et al. 2004). In addition, such logic is perceived to follow human-based reasoning to the extent that they are parameterized by means of rules consisting of linguistic expressions. Fuzzy systems, however, are not void of

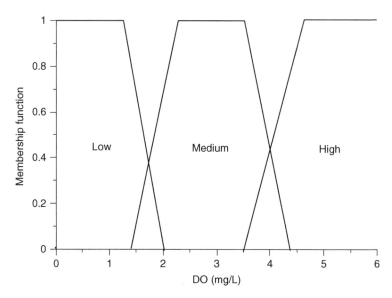

FIGURE 6.21   Generalized set membership function (*y*-axis) of dissolved oxygen (DO).

problems. For example, such logic is not appropriate to deal with the imprecision associated with datasets prone to large amounts of missing data (Sproule et al. 2002).

6.12   EXPANDED RESEARCH APPLICATION—A MULTIVARIATE STUDY OF THE RELATIONS BETWEEN PM₁₀ COMPOSITION AND CELL TOXICITY

Section 1.3.1 discusses the importance of being able to measure and quantify particulate matter (PM) concentrations in polluted air successfully because of the deleterious health effects that are associated with exposure. The importance of particle mass and size (e.g., $PM_{2.5}$ and PM_{10}) in linking PM to the harmful effects is also stated. Studies have shown, however, that that such effects are observed even when mass concentrations are within acceptable levels (Samet et al. 2000). In addition, recent studies have shown that particle composition, and even geographical location, are playing important roles in PM toxicity (Samet et al. 2000, Roger et al. 2005). The following detailed examination by Pérez and colleagues critically evaluates and further explains the relations between PM_{10} composition and cell toxicity with the aid of multivariate statistical techniques (Pérez et al. 2007).

6.12.1   Statement of Problem

The sources given earlier have provided evidence suggesting that particulate composition and geographical location play roles in PM toxicity; however, little is known about how these components specifically participate (individually or in groups) in the toxicity and specific outcomes at the cellular level.

6.12.2 Research Objectives

The research objectives of this study were to utilize multivariate and graphical methods to explore the relationships among PM_{10} components in order to elucidate their covariant structure and explore their relationships with cell toxicity at three geographical locations in Mexico City.

6.12.3 Experimental Methods and Calculations

Samples were taken at three distinct locations in Mexico City: Iztacala (North—manufacturing district), Merced (Central—heavy traffic region), and the Ciudad Universitaria (South—residential area) using PM_{10} samplers. A variety of methods were utilized to analyze particulate matter for concentration (gravimetric analysis), elemental composition (particle induced X-ray emission), carbon content analysis (gravimetric analysis performed with carbon content determined using the evolved gas technique), endotoxin determination [kinetic chromagenic limulus amebocyte lysate (LAL) assay] and biological responses including inhibition of cell prolife (ration, Interleukin-6 (IL-6), tumor necrosis factor-alpha (TNFα), and protein 53 (p53) enzyme-linked immuno sorbent assay (ELISA)].

Concerning cell toxicity, IL-6 is a pro-inflammatory cytokine secreted by T cells and macrophages to stimulate immune response. TNFα is a cytokine involved in apoptotic cell death, cellular proliferation, differentiation, inflammation, tumori-genesis, and viral replication. Finally, p53 is a transcription factor that regulates the cell cycle, and is also known to function as a tumor suppressor.

A variety of statistical techniques were employed to assess PM components [exploratory principal component analysis, one-way ANOVA, Levene's test (for validating equal variances), and Tukey's method] as multiple comparison tests for determining differences among sites. A factorial design was utilized to evaluate the effect of site and PM_{10} concentrations on the biological responses (inhibition of cell proliferation, IL-6, and TNFα). Here, the Tukey test was applied for determining subgroups of main effects when interaction effects were noticeably absent. A Welch test was used by the investigators to determine differences in p53 among sites with the incorporation of the Games–Howell approach as a multiple comparison post-test. Finally, a radial plot was used in order to explore patterns of association between PM_{10} composition and cell toxicity.

6.12.4 Results and Interpretation

Overall, results confirmed heterogeneity in particle mass, composition, and toxicity in samples collected at the three locations. Table 6.23 lists the descriptive statistics of PM_{10} components (μg/mg) from the three sites. As shown, elemental masses were generally smaller in the south with both carbon fractions having lower levels

TABLE 6.23

Descriptive Statistics of PM$_{10}$ Components (μg/mg) in Samples from the Three Mexico City Sites

Component	Statistic	North	Central	South
S	Mean	71.562	84.338	16.807
	Std. deviation	31.375	22.951	16.947
	Minimum	31.754	41.005	6.238
	Maximum	126.989	122.172	63.195
Cl	Mean	5.771	8.695	5.448
	Std. deviation	1.019	2.101	3.442
	Minimum	4.311	6.855	2.662
	Maximum	7.388	12.358	13.057
K	Mean	16.717	24.894	6.358
	Std. deviation	5.147	4.743	2.408
	Minimum	11.250	18.670	3.322
	Maximum	26.500	31.587	10.829
Ca	Mean	52.606	70.837	41.914
	Std. deviation	15.050	11.433	15.182
	Minimum	35.127	56.519	23.340
	Maximum	81.214	91.864	72.258
Ti	Mean	3.771	4.725	3.001
	Std. deviation	0.526	7.749	0.965
	Minimum	2.987	3.118	1.439
	Maximum	4.450	8.592	4.396
V	Mean	0.830	0.577	0.377
	Std. deviation	0.935	0.243	0.308
	Minimum	0.152	0.325	0.006
	Maximum	2.862	1.109	0.915
Cr	Mean	0.487	0.666	0.676
	Std. deviation	0.257	0.252	0.612
	Minimum	0.159	0.312	0.112
	Maximum	1.036	1.022	2.107
Mn	Mean	1.304	1.571	0.727
	Std. deviation	0.416	0.317	0.234
	Minimum	0.955	1.177	0.321
	Maximum	2.262	1.988	1.070
Fe	Mean	36.001	44.017	23.282
	Std. deviation	6.395	13.988	7.362
	Minimum	27.802	27.320	12.195
	Maximum	46.133	64.641	36.761
Ni	Mean	0.172	0.145	0.158
	Std. deviation	0.126	0.119	0.213
	Minimum	0.000	0.053	0.000
	Maximum	0.409	0.421	0.728

continued

TABLE 6.23   (continued)

Component	Statistic	North	Central	South
Cu	Mean	0.442	1.323	0.412
	Std. deviation	0.151	0.405	0.728
	Maximum	0.209	0.835	0.050
	Minimum	0.665	1.944	2.462
Zn	Mean	5.778	6.704	0.928
	Std. deviation	3.208	2.504	1.134
	Minimum	1.431	3.866	0.268
	Maximum	11.107	9.615	4.126
Pb	Mean	2.828	2.535	0.024
	Std. deviation	1.144	1.782	0.074
	Minimum	1.534	1.211	0.000
	Maximum	4.600	6.788	0.235
Endotoxins	Mean	0.006	0.014	0.014
	Std. deviation	0.004	0.005	0.009
	Minimum	0.001	0.006	0.005
	Maximum	0.011	0.019	0.034
Organic carbon	Mean	257.613	357.475	339.040
	Std. deviation	82.930	84.316	55.384
	Minimum	169.400	231.900	255.500
	Maximum	398.100	490.800	440.500
Elemental carbon	Mean	51.100	83.788	73.060
	Std. deviation	16.938	10.447	41.011
	Minimum	28.600	73.100	39.100
	Maximum	83.900	101.100	151.800

Source: From Pérez, I. R., et al., *Chemosphere*, 67, 1218–1228, 2007. With permission from Elsevier.

in the northern site. In addition, associated endotoxins had larger concentrations in the center and south portions of Mexico City. Table 6.24 provides the matrix correlation of PM_{10} components with significance [$p < 0.01$; t-test with Bonferroni correction (Afifi and Clark 1990)]. Principal component analysis was used to reduce the dimensions of the overall dataset with three extracted PCs: Group 1—S/K/Ca/Ti/Mn/Fe/Zn/Pb, Group 2—Cl/Cr/Ni/Cu, and Group 3—endotoxins/OC/EC explaining 73% of the variance. Evaluation of the PCs by site showed that the average of Group 1 were statistically different among north, center, and south locations with the central site producing the highest values (Figure 6.22a). The averages for Group 2 were similar among sites (Figure 6.22b) with Group 3 having lower averages in the north compared with the other two sties (Figure 6.22c). These results were supported by previous work performed by the investigators (Vega et al. 2004). Finally, a radial plot (Figure 6.23) shows the relationship between PM_{10} component groups and biological effects stratified by sampling site. A radial plot is a graphical display of radial lines, polygons, or symbols centered at a midpoint of the plot frame, with links or positions corresponding to the magnitude of

TABLE 6.24

Matrix Correlation of PM$_{10}$ Components

	S	Cl	K	Ca	Ti	V	Cr	Mn	Fe	Ni	Cu	Zn	Pb	Endotoxins	Organic Carbon	Elemental Carbon
S	1.00	0.41	0.79	0.58	0.35	0.31	0.11	0.65	0.53	0.11	0.50	0.66	0.55	-0.20	-0.14	0.00
Cl		1.00	0.49	0.73	0.69	0.16	0.66	0.50	0.67	0.51	0.77	0.44	0.18	0.28	-0.10	0.15
K			1.00	0.80	0.67	0.20	-0.10	0.80	0.84	-0.03	0.42	0.68	0.65	-0.07	-0.02	0.12
Ca				1.00	0.68	0.39	0.25	0.75	0.83	0.19	0.50	0.63	0.50	0.22	-0.05	0.18
Ti					1.00	0.17	0.14	0.61	0.89	0.38	0.52	0.46	0.33	0.04	0.01	0.08
V						1.00	0.25	0.33	0.27	0.53	0.16	0.50	0.27	-0.17	-0.14	-0.13
Cr							1.00	0.14	0.05	0.62	0.64	0.13	-0.17	0.32	-0.07	-0.08
Mn								1.00	0.80	0.20	0.54	0.83	0.68	-0.10	0.01	0.00
Fe									1.00	0.23	0.45	0.66	0.54	-0.01	-0.06	0.10
Ni										1.00	0.57	0.35	0.03	-0.19	-0.13	-0.23
Cu											1.00	0.51	0.17	0.06	0.11	0.07
Zn												1.00	0.78	-0.27	-0.12	-0.11
Pb													1.00	-0.21	-0.10	-0.17
Endotoxins														1.00	0.30	0.52
Organic Carbon															1.00	0.47
Elemental Carbon																1.00

Source: Modified from Pérez, I. R., et al., *Chemosphere*, 67, 1218–1228, 2007. With permission from Elsevier.

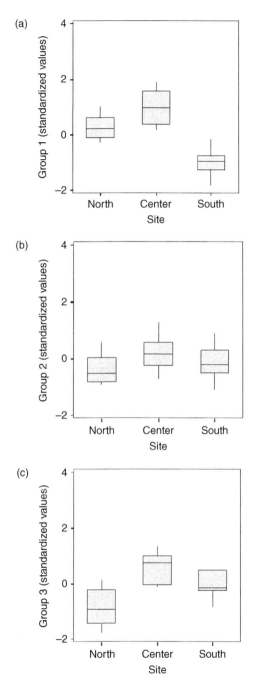

FIGURE 6.22 Multiple box plots of the three principal components with: (a) Group 1, (b) Group 2, and (c) Group 3. Box plots represent the median and the 25th and 75th percentile values with lines indicating minimum and maximum values. Outlier values are represented by an asterisk. (From Pérez, I. R., et al., *Chemosphere*, 67, 1218–1228, 2007. With permission from Elsevier.)

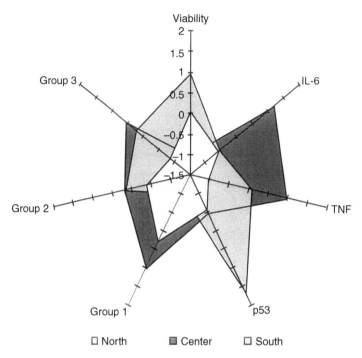

FIGURE 6.23 Radial plot showing the relationship between PM10 component groups (stratified by sampling site) and biological effects. (From Pérez, I. R., et al., *Chemosphere*, 67, 1218–1228, 2007. With permission from Elsevier.)

the numeric data values (Galbraith 1994). As shown in Figure 6.23, the existence of different pollutant mixtures at each site related to different biological effects can be identified.

6.12.5 SUMMARY AND SIGNIFICANCE

Multivariate statistical analysis proved highly beneficial in demonstrating that PM_{10} samples from three different locations in Mexico City consisted of differing compositions and produces a variety of biological (cellular) effects. The investigators were successful in identifying three independent components via PCA that vary in concentrations in the three sites studied.

6.13   CHAPTER SUMMARY

The aim of this chapter is to provide an introduction to multivariate data analysis techniques and to show how these can be effectively used to understand and characterize complex environmental datasets. As noted, data preprocessing is an important step in multivariate analysis with obvious implications in enhancing information and providing improved robustness. Several techniques are covered including correlation analysis, factor analytic techniques, pattern recognition techniques, multivariate calibration, and soft computing techniques. Although not all-encompassing, a wide variety of techniques are presented, highlighting a multitude of applications. This developing field will no doubt play a major role in shaping environmental analysis for years to come.

6.14   END OF CHAPTER PROBLEMS

1. An instructor is interested in determining if it is possible to predict the score of a chemometrics exam from the amount of time spent studying for the exam. Choose the proper explanatory variable in this study:
   (a) The score of the exam.
   (b) The instructor.
   (c) The amount of time spent studying.
   (d) The difficulty of the exam.
2. In terms of the correlation coefficient, r, which of the following is true?
   (a) $-1 \le r \le 1$.
   (b) There is no association among variables.
   (c) If r is the correlation between X and Y, then $-r$ is the correlation between Y and X.
   (d) All of the above.
   (e) None of the above.
3 Using your chemometrics/statistical package of choice, produce a correlation matrix for the data presented in Table 6.10.
4. Describe the key differences between factor analysis and principal component analysis.
5. Using your chemometrics/statistical package of choice:
   (a) Produce a dendogram of Table 6.10 via CA employing Ward's method.
   (b) Do the environmental parameters fall in groups?
6. List the advantages of multivariate calibration techniques over the traditional univariate approach. What are potential problems with the multivariate calibration approach?
7. Identify aspects of ANN that are superior to other methods exposing underlying relationships in environmental datasets.

REFERENCES

Absalan, G. and Nekoeinia, M. 2005. Simultaneous kinetic determination of Fe(II) and Fe(III) based on their reactions with NQT4S in micellar media by using PLS and PCR methods. *Analytica Chimica Acta* 531: 293–298.

Adriaenssens, V., De Baets, B., Goethals, L. M., and De Pauw, N. 2004. Fuzzy rule-based models for decision support in ecosystem management. *The Science of the Total Environment* 319: 1–12.

Afifi, A. A. and Clark, V. 1990. *Computer-aided Multivariate Analysis.* New York: Van Nostrand Reinhold Company.

Al-Degs, Y. S., El-Sheikh, A. H., Al-Ghouti, M. A., Hemmateenejad, B., and Walker, G. M. 2008. Solid-phase extraction and simultaneous determination of trace amounts of sulphonated and azo sulphonated dyes using microemulsion-modified-zeolite and multivariate calibration. *Talanta* 75: 904–915.

Alsberg, B. K., Goodacre, R., Rowland, J. J., and Kell, D. B. 1997. Classification of pyrolysis mass spectra by fuzzy multivariate rule induction—comparison with regression, K-nearest neighbour, neural and decision-tree methods. *Analytica Chimica Acta* 348: 389–407.

Amaral, A. L. and Ferreira, E. C. 2005. Activated sludge monitoring of a wastewater treatment plant using image analysis and partial least squared regression. *Analytica Chemica Acta* 544: 246–253.

Amini, N. and McKelvie, I. D. 2004. An enzymatic flow analysis method for the determination of phosphatidylcholine in sediment pore waters and extracts. *Talanta* 66: 445–452.

Astel, A. 2007. Chemometrics based on fuzzy logic principles in environmental studies. *Talanta* 72: 1–12.

Azzouz, T. and Tauler, R. 2008. Application of multivariate curve resolution alternating least squares (MCR-ALS) to the quantitative analysis of pharmaceutical and agricultural samples. *Talanta* 74: 1201–1210.

Basheer, I. A. and Hajmeer, M. 2000. Artificial neural networks: fundamentals, computing, design and application. *Journal of Microbiological Methods* 43: 3–31.

Berrueta, L. A., Alonso-Salces, R. M., and Héberger, K. 2007. Supervised pattern recognition in food analysis. *Journal of Chromatography A* 1158: 196–214.

Bock, H. H. 1985. On significance tests in cluster analysis. *Journal of Classification* 2: 77–108.

Chatterjee, M., Silva Filho, E. V., Sarkar, S. K., Sella, S. M., Bhattacharya, A., Satpathy, K. K., Prasad, M. V. R., Chakraborty, S., and Bhattacharya, B. D. 2007. Distribution and possible source of trace elements in the sediment cores of a tropical macrotidal estuary and their ecotoxicolocial significance. *Environment International* 33: 346–356.

Conway, J. M. and Huffcutt, A. I. 2003. A review and evaluation of exploratory factor analysis practices in organizational research. *Organizational Research Methods* 6: 147–168.

Deconinck, E., Coomans, D., and Vander Heyden, Y. 2007. Exploration of linear modelling techniques and their combination with multivariate adaptive regression splines to predict gastro-intestinal absorption of drugs. *Journal of Pharmaceutical and Biomedical Analysis* 43: 119–130.

Estienne, F., Massart, D. L., Zanier-Szydlowski, N., and Marteau, P. 2000. Multivariate calibration with Raman spectroscopic data: a case study. *Analytica Chimica Acta* 424: 185–201.

Fraley, C. and Raftery, A. E. 2002. Model-based clustering, discriminant analysis, and density estimation. *Journal of the American Statistical Association* 97: 611–631.

Galbraith, R. F. 1994. Some applications of radial plots. *Journal of the American Statistical Association* 89: 1232–1242.

Garrido Frenich, A., Martínez Galera, M., Martínez Vidal, J. L., and Gil García, M. D. 1996. Partial least-squares and principal component regression of multi-analyte high-performance liquid chromatography with diode-array detection. *Journal of Chromatography A* 727: 27–38.

Gong, X. and Richman, M. B. 1995. On the application of cluster analysis to growing season precipitation data in North America East of the Rockies. *Journal of Climate* 8: 897–931.

Hartigan, J. A. and Wong, M. A. 1979. Algorithm AS 136: a K-means clustering algorithm. *Applied Statistics* 28: 100–108.

Ho, C. K. and Hughes, R. C. 2002. In situ chemiresistor sensor package for realtime detection of volatile organic compounds in soil and groundwater. *Sensors* 2: 23–24.

Hoeffding, W. 1948. A non-parametric test of independence. *Annals of Mathematics and Statistics* 19: 546–557.

Hopfield, J. J. 1982. Neural networks and physical systems with emergent collective computational abilities. *Proceedings of the National Academy of Sciences* 79: 2554–2558.

Huitao, L., Ketai, W., Hongping, X., Xingguo, C., and Zhide, H. 2002. Application of experimental design and artificial neural networks for separation and determination of active components in traditional Chinese medicinal preparations by capillary electrophoresis. *Chromatographia* 55: 579–583.

Hupp, A. M., Marshall, L. J., Campbell, D. I., Smith, W., and McGuffin, V. L. 2008. Chemometric analysis of diesel fuel for forensic and environmental applications. *Analytica Chimica Acta* 606: 159–171.

Liu, J., Chen, S., and Tan, X. 2008. A study on three linear discriminant analysis based methods in small sample size problems. *Pattern Recognition* 41: 102–116.

JMP. 2007. Statistics and Graphics Guide. SAS Institute: Cary, North Carolina.

Kassomenos, P., Karakitsios, S., and Papaloukas, C. 2006. Estimation of daily traffic emissions in a South-European urban agglomeration during a workday. Evaluation of several "what if" scenarios. *Science of the Total Environment* 370: 480–490.

Koehler, J., Hilt, S., Adrian, R., Nicklisch, A., Kozerski, H. P., and Walz, N. 2005. Long-term response of a shallow, moderately flushed lake to reduced external phosphorus and nitrogen loading. *Freshwater Biology* 50: 1639–1650.

Kopp, B. 1978. Hierarchical classification 1: single-linkage method. *Biometrical Journal* 20: 495–501.

Kohonen, T. 1997. *Self-Organizing Maps*. Springer-Verlag: New York.

Knöscke, R. 2005. Organic sediment nutrient concentrations and their relationship with the hydrological connectivity of flood plain waters. *Hydrobiologia* 560: 63–76.

Laranjeira, R., Pillon, S., and Dunn, J. 2000. Environmental tobacco smoke exposure among non-smoking waiters: measurement of expired carbon monoxide levels. *Sao Paulo Medical Journal* 118: 89–92.

Lobscheid, A., McKone, T. E., and Vallero, D. A. 2007. Exploring relationships between outdoor air particulate-associated polycyclic aromatic hydrocarbon and $PM_{2.5}$: a case study of benzo(a)pyrene in California metropolitan regions. 2007. *Atmospheric Environment* 41: 5659–5672.

Lundstedt-Enkel, K., Asplund, L., Nylund, K., Bignert, A., Tysklind, M., Olsson M., and Örberg, J. 2006. Multivariate data analysis of organochlorines and brominated flame retardants in Baltic Sea guillemot (*Uria aalge*) egg and muscle. *Chemosphere* 65: 1591–1599.

Manera, M., Miró, M., Estels, J. M., and Cerdá, V. 2007. Multi-syringe flow injection solid-phase extraction system for on-line simultaneous spectrophotometric determination of nitro-substituted phenol isomers. *Analytica Chimica Acta* 582: 41–49.

Martín, M. L., Turias, I. J., González, F. J., Galindo, P. L., Trujillo, F. J., Puntonet, C. G., and Gorriz, J. M. 2008. Prediction of CO maximum ground level concentrations in the Bay of Algeciras, Spain using artificial neural networks. *Chemosphere* 70: 1190–1195.

Mason, R. L. and Young, J. C. 2005. Multivariate tools: principal component analysis. *Quality Progress* 38: 83–85.

Meklin, T., Reporen, T., McKinstry, C., Cho, S-H., Grinshpun, S. A., Nevalainen, A., Vepsäläinen, A., Hangland, R. A., LeMasters, G., and Vesper, S. J. 2007. Comparison of mold concentrations quantified by MSQPCR in indoor and outdoor air sampled simultaneously. *Science of the Total Environment* 382: 130–134.

Miller, J. N. and Miller, J. C. 2000. *Statistics and Chemometrics for Analytical Chemistry*. Prentice Hall: Essex.

Milligan, G. W. 1980. An examination of the effect of six types of error perturbation on fifteen clustering algorithms. *Psychometrika* 45: 325–342.

Meissl, K., Smidt, E., and Schwanninger, M. 2007. Prediction of humic acid content and respiration activity of biogenic waste by means of Fourier transform infrared (FTIR) spectra and partial least squares regression (PLS-R) models. *Talanta* 72: 791–799.

Ocampo-Duque, W., Ferré-Huguet, N., Domingo, J. L., and Schumacher, M. 2006. Assessing water quality in rivers with fuzzy inference systems: a case study. *Environment International* 32: 733–742.

Otto, M. 1999. *Chemometrics*. Weinheim: Wiley-VCH.

Pedrycz, W. and Gomide, F. 1998. *An Introduction to Fuzzy Sets: Analysis and Design Complex Adaptive Systems*. Cambridge: MIT Press.

Pérez, I. R., Serrano, J., Alfaro-Moreno, E., Baumgardner, D., García-Cuellar, del Campo, J. M. M., Raga, G. B., Castillejos, M., Colín, R. D., and Vargas, A. R. O. 2007. Relations between PM_{10} composition and cell toxicity: a multivariate and graphical approach. *Chemosphere* 67: 1218–1228.

Rivera, D., Alam, M. K., Davis, C. E., and Ho, C. K. 2003. Characterization of the ability of polymeric chemiresistor arrays to quantitate trichloroethylene using partial least squares (PLS): effects of experimental design, humidity, and temperature. *Sensors and Actuators B* 92: 110–120.

Robotti, E., Bobba, M., Panepinto, A., and Marengo, E. 2007. Monitoring of the surface of paper samples exposed to UV light by ATR-FT-IR spectroscopy and use of multivariate control charts. *Analytical and Bioanalytical Chemistry* 388: 1249–1263.

Rodriguez, I., Bollain, M. H., and Cela, R. 1996. Quantifications of two chromatographic unresolved dichlorophenols using gas chromatography-direct deposition-Fourier transform infrared spectrometry and multivariate calibration. *Journal of Chromatography* 750: 341–349.

Roger, D., Peng, R. D., Dominici, F., Pastor-Barriuso, R., Zeger, S. L., and Samet, J. M. 2005. Seasonal analysis of air pollution and mortality in 100 US cities. *American Journal of Epidemiology* 161: 585–594.

Roggo, Y., Dupounchel, L., and Huvenne, J. P. 2003. Comparison of supervised pattern recognition methods with McNemar's statistical test. Application to quantitative analysis of sugar beat by near-infrared spectroscopy. *Analytica Chimica Acta* 477: 187–200.

Rumelhart, D., Hinton, G. E., and Williams, R. J. 1986. *Parallel Distributed Processing: Explorations in the Microstructure of Cognition*. Cambridge, MA: MIT Press.

Rupinski, M. T. and Dunlap, W. P. 1996. Approximating Pearson product-moment correlations from Kendall's Tau and Spearman's Rho. *Educational and Psychological Measurement* 56: 419–429.

Samarasinghe, S. 2006. *Neural Networks for Applied Sciences and Engineering: From Fundamentals to Complex Pattern Recognition*. Boca Raton: Auerbach Publications.

Samet, J. M., Dominici, F., Curriero, F. C., Coursac, I., and Zeger, S. L. 2000. Fine particulate air pollution and mortality in 20 US cities, 1987–1994. *New England Journal of Medicine* 343: 1742–1749.

Saridakis, K. M. and Dentsoras, A. J. 2008. Soft computing in engineering design—a review. *Advanced Engineering Informatics* 22: 201–221.

Spears, B. M., Carvalho, L., Perkins, R., Kirika, A., and Paterson, D. M. 2007. Sediment phosphorus cycling in a large shallow lake: Spatio-temporal variation in phosphorus pools and release. *Hydrobiologia* 584: 37–48.

Sproule, B. A., Naranjo, C. A., and Türksen, I. B. 2002. Fuzzy pharmacology: theory and applications. *Trends in Pharmacological Sciences* 23: 412–417.

Tan, H. and Brown, S. D. 2003. Multivariate calibration of spectral data using dual-domain regression analysis. *Analytica Chimica Acta* 490: 291–301.

Thomas, E. V. and Ge, N. X. 2000. Development of robust multivariate calibration models. *Technometrics* 42: 168–177.

Trevisan, G. V. and Forsberg, B. R. 2007. Relationships among nitrogen and total phosphorus, algal biomass and zooplankton density in the central Amazonia lakes. *Hydrobiologia* 586: 357–365.

Upadhaya, B. and Eryureka, E. 1992. Application of neural network for sensory validation and plant monitoring. *Neural Technology* 97: 170–176.

Vega, E., Reyes, E., Ruiz, H., García, J., Sánchez, G., Martínez-Villa, G., González, U., Chow, J. C., and Watson, J. G. 2004. Analysis of $PM_{2.5}$ and PM_{10} in the atmosphere of Mexico City during 2000–2002. *Journal of the Air and Waste Management Association* 54: 786–798.

Verdú-Andrés, J., Massart, D. L., Menardo, C., and Sterna, C. 1999. Correction of non-linearities in spectroscopic multivariate calibration by using transformed original variables. Part II. Application to principal component regression. *Analytica Chimica Acta* 389: 115–130.

Vitale, M., Anselmi, S., Salvatori, E., and Manes, F. 2007. New approaches to study the relationship between stomal conductance and environmental factors under Mediterranean climatic conditions. *Atmospheric Environment* 41: 5385–5397.

Walmsley, A. D. 1997. Improved variable selection procedure for multivariate linear regression. *Analytica Chimica Acta* 354: 225–232.

Widrow, B. and Lehr, M. 1990. 30 years of adaptive neural networks: perception, madaline and back propagation. *Proceedings of the IEEE* 78: 1415–1451.

Wold, S. 1978. Cross-validatory estimation of the number of components in factor and principal components models. *Technometrics* 20: 397–405.

Wold, S., Sjöstöm, M., Carlson, R., Lundstedt, T., Hallberg, S., Skagerberg, B., Wikstrom, C., and Öhman, J. 1986. Multivariate design. *Analytica Chimica Acta* 191: 17–32.

Wu, K.-S., Huo, X., and Zhu, G.-H. 2008. Relationships between esophageal cancer and spatial environmental factors by using Geographical Information System. *Science of the Total Environment* 393: 219–225.

Wunderlin, D. A., Diaz, M. P., Ame, M. V., Pesce, S. F., Hued, A. C., and Bistomi, M. A. 2001. Pattern recognition techniques for the evaluation of spatial and temporal variations in water quality. A case study: Suquía River basin (Córdoba–Argentina). *Water Research* 35: 2881–2894.

Zadeh, L. A. 1965. Fuzzy sets. *Inference Control* 8: 338–353.

Appendix I
Common Excel®
Shortcuts and Key
Combinations

Shortcuts and Combination Keys	Function
F1	Excel® help is opened
F2	Activates a given cell
F3	Activates the Paste command
F4	Repeats most recent worksheet action
F5	Opens the Go To dialog box
F6	Moves to the next pane
F7	Spell check is activated
F8	Extends a given selection
F9	Calculates all worksheets in all open workbooks
F10	Selects the Menu bar
F11	Creates a chart of the current data
F12	Opens the Save As dialog box
Shift + F2	Edits a cell comment
Shift + F3	Opens the Function dialog box
Shift + F5	Activates the Find command
Shift + F6	Switches to the previous pane in a worksheet
Shift + F8	Enables the addition of nonadjacent cells to another selection of cells
Shift + F9	Calculates the entire worksheet
Shift + F10	Displays the Shortcut menu for the selected item
Shift + F12	Saves the active worksheet
Shift + Ins	Pastes the data from the clipboard
Shift + Home	Selects the cells from the active cell to the left edge of the current row
Shift + PgUp	Selects the cells from the active cell to the top of the current column
Ctrl + F1	Closes and reopens the current task pane

Ctrl + F3	Opens the Define Name dialog box
Ctrl + F4	Closes the workbook window
Alt + D	Displays the Data menu
Alt + E	Displays the Edit menu
Alt + F	Displays the File menu
Alt + H	Displays the Help menu
Alt + I	Displays the Insert menu
Alt + O	Displays the Format menu
Alt + T	Displays the Tools menu
Alt + V	Displays the View menu
Alt + W	Displays the Windows menu

Appendix II
Symbols Used in Escuder-Gilabert et al. (2007)

Symbol	Meaning
α	Level of significance (type I error)
β	β-error (type II error)
CL	Confidence level
E	Estimated relative error for μ
E_0	Accepted true relative error
E_{LIM}	Limit for the relative error in trueness assessment
$e_{i,j}$	Random error under repeatability conditions for the ith replicate and jth run
f_i	Random run effect for the jth run
i	Index of replicates (from 1 to Nr)
j	Index of runs (from 1 to Ns)
$\mu \pm s_\mu$	Experimental grand mean and standard deviation corresponding to the vector of j means
μ_0	Accepted true value for μ
Nr	Number of replicates
Ns	Number of runs
$RSDr$	Relative standard deviation under repeatability conditions
$RSDrun$	Between run relative standard deviation
$RSDi$	Relative standard deviation under intermediate precision conditions
$RSDr_0$	True $RSDr$
$RSDrun_0$	True $RSDrun$
$RSDi_0$	True $RSDi$
$RSDi_{\mathrm{LIM}}$	Limit for the $RSDi$ in precision assessment
σ_r^2	Variance under repeatability conditions
σ_{run}^2	Between-run variance
$\mathbf{X}$	Matrix of size $Nr \times Ns$
$\mathbf{x}_j$	Vector of replicate values under given conditions (run j)
x_{ij}	Data related to the ith replicate of the jth run
$\pm U$	Estimated expanded uncertainty interval

Appendix III
Review of Basic Matrix Algebra Notation and Operations

A matrix is a table of numbers consisting of n rows and m columns (i.e., a $n \times m$ matrix):

$$\mathbf{A} = \begin{pmatrix} a_{11} & a_{12} & a_{1m} \\ a_{21} & a_{22} & a_{2m} \\ \vdots & & \vdots \\ a_{n1} & a_{n2} & a_{nm} \end{pmatrix} \text{ or with real numbers: } \mathbf{A} = \begin{pmatrix} 1 & 4 & 6 \\ 2 & 9 & 7 \\ 3 & 5 & 8 \end{pmatrix}$$

where an individual element of $\mathbf{A} = a_{ij}$. The first subscript in a matrix refers to the row and the second to the column. A square matrix consists of the same number of rows and columns (i.e., $n \times n$ matrix). Note that matrix $\mathbf{A}$ here is square but matrix $\mathbf{B}$ below is not:

$$\mathbf{B} = \begin{pmatrix} 1 & 4 & 6 \\ 2 & 9 & 7 \end{pmatrix}$$

A vector is a type of matrix that has only one row (i.e., a row vector) or one column (i.e., a column vector). In the following, $\mathbf{a}$ is a column vector while $\mathbf{b}$ is a row vector.

$$a = \begin{pmatrix} 6 \\ 2 \end{pmatrix} \qquad b = (3 \quad 1 \quad 6)$$

A symmetric matrix is a square matrix in which $a_{ij} = a_{ji}$ for all i and j and is presented as

$$A = \begin{pmatrix} 3 & 4 & 1 \\ 4 & 8 & 5 \\ 1 & 5 & 0 \end{pmatrix}$$

Do not confuse this with a diagonal matrix, a symmetric matrix where all the off-diagonal elements are 0:

$$D = \begin{pmatrix} 3 & 0 & 0 \\ 0 & 1 & 0 \\ 0 & 0 & 5 \end{pmatrix}$$

An identity matrix is a diagonal matrix with 1s on the diagonal:

$$I = \begin{pmatrix} 1 & 0 & 0 \\ 0 & 1 & 0 \\ 0 & 0 & 1 \end{pmatrix}$$

In a transposed matrix A', the rows and columns are interchanged as in the following. Compare that to the first matrix presented earlier.

$$A = \begin{pmatrix} a_{11} & a_{21} & a_{n1} \\ a_{12} & a_{22} & a_{n2} \\ \vdots & & \vdots \\ a_{1m} & a_{2m} & a_{nm} \end{pmatrix} \text{ or with real numbers: } A = \begin{pmatrix} 1 & 2 & 3 \\ 4 & 9 & 5 \\ 6 & 7 & 8 \end{pmatrix}$$

For matrix addition, each element of the first matrix is added to the corresponding element of the second to produce a result. Note that the two matrices must have the same number of rows and columns:

$$\begin{pmatrix} 0 & 2 & -1 \\ 3 & 1 & 3 \\ 1 & 6 & 0 \end{pmatrix} + \begin{pmatrix} 10 & -1 & 3 \\ 2 & 0 & 2 \\ 3 & 1 & 1 \end{pmatrix} = \begin{pmatrix} 10 & 1 & 2 \\ 5 & 1 & 5 \\ 4 & 7 & 1 \end{pmatrix}$$

Note that matrix subtraction works in the same way, except that elements are subtracted instead of added.

Multiplication of an $n \times n$ matrix $\mathbf{A}$ and an $n \times n$ matrix $\mathbf{B}$ given a result of $n \times n$ matrix $\mathbf{C}$:

$$\begin{pmatrix} 2 & 1 & 3 \\ -2 & 2 & 1 \end{pmatrix} \begin{pmatrix} 2 & 1 \\ 3 & 2 \\ -2 & 2 \end{pmatrix} = \begin{pmatrix} 1 & 10 \\ 0 & 4 \end{pmatrix}$$

Note that in multiplication, $\mathbf{A} \times \mathbf{B}$ does not generally equal $\mathbf{B} \times \mathbf{A}$. In other words, matrix multiplication is not commutative. Considering associative and distributive laws: $(\mathbf{A} \times \mathbf{B}) \times \mathbf{C} = \mathbf{A} \times (\mathbf{B} \times \mathbf{C})$; $\mathbf{A} \times (\mathbf{B} + \mathbf{C}) = \mathbf{A} \times \mathbf{B} + \mathbf{A} \times \mathbf{C}$ and $(\mathbf{B} + \mathbf{C}) \times \mathbf{A} = \mathbf{B} \times \mathbf{A} + \mathbf{C} \times \mathbf{A}$.

Typically, one takes the multiplication by an inverse matrix as the equivalent of matrix division. The inverse of a matrix is that matrix which, when multiplied by the original matrix, gives an identity ($\mathbf{I}$) matrix with the inverse of denoted by a superscripted -1:

$$\mathbf{A}^{-1}\mathbf{A} = \mathbf{A}\mathbf{A}^{-1} = \mathbf{I}$$

Note that to have an inverse, a matrix must be square. Consider the following matrix:

$$\mathbf{A} = \begin{pmatrix} a_{11} & a_{12} \\ a_{21} & a_{22} \end{pmatrix}$$

The inverse of this matrix exists if $a_{11}a_{22} - a_{12}a_{21} \neq 0$. If the inverse exists, it is then given by

$$\mathbf{A} = \frac{1}{a_{11}a_{22} - a_{12}a_{21}} \begin{pmatrix} a_{22} & -a_{12} \\ -a_{21} & a_{11} \end{pmatrix}$$

Note that for covariance and correlation matrices, those encountered in Chapter 6, an inverse will always exist, provided that there are more subjects than there are variables and that every variable has a variance greater than 0. The existence of the inverse is dependent upon the determinant, a scalar-values function of the matrix. For example

$$\det \mathbf{A} = \begin{pmatrix} a_{11} & a_{12} \\ a_{21} & a_{22} \end{pmatrix} = a_{11}a_{22} - a_{12}a_{21}$$

For covariance and correlation matrices, the determinant is a number that often expresses the generalized variance of the matrix. Here, covariance matrices with small determinants denote variables that are highly correlated (see discussions on factor analysis or regression analysis).

Defining the determinant can help in formalizing the general form of the inverse matrix:

$$\mathbf{A}^{-1} = \frac{1}{\det \mathbf{A}} \text{adj } \mathbf{A}$$

where adj $\mathbf{A}$ = the adjugate of $\mathbf{A}$. There are a couple of ways to compute an inverse matrix, with the easiest typically being in the form of an augmented matrix $(\mathbf{A}|\mathbf{I})$ from $\mathbf{A}$ and $\mathbf{I}_n$, and then utilizing Gaussian elimination to transform the left half into $\mathbf{I}$. Once completed, the right half of the augmented matrix will be $\mathbf{A}^{-1}$. Additionally, one can compute the i,jth element of the inverse by using the general formula

$$\mathbf{A}_{ji} = \frac{C_{ij}\mathbf{A}}{\det \mathbf{A}}$$

where $C_{ij} = i, j$th cofactor expansion of matrix $\mathbf{A}$.

An orthogonal matrix has the general form: $\mathbf{AA}^t = \mathbf{I}$. Thus, the inverse of an orthogonal matrix is simply the transpose of that matrix. Orthogonal matrices are very important in factor analysis. Matrices of eigenvectors (discussed in Chapter 6) are orthogonal matrices. Note that only square matrices can be orthogonal matrices. As discussed, eigenvalues and eigenvectors of a matrix play an important part in multivariate analysis. General concepts regarding eigenvalues and eigenvectors include the following:

1. Eigenvectors are scaled so that $\mathbf{A}$ is an orthogonal matrix;
2. An eigenvector of a linear transformation is a nonzero vector that is either left unaffected or simply multiplied by a scale factor after transformation;
3. The eigenvalue of a nonzero eigenvector is the scale factor by which it has been multiplied;
4. An eigenvalue reveals the proportion of total variability in a matrix associated with its corresponding eigenvector;
5. For a covariance matrix, the sum of the diagonal elements of the covariance matrix equals the sum of the eigenvalues;
6. For a correlation matrix, all the eigenvalues sum to n, the number of variables;
7. The decomposition of a matrix into relevant eigenvalues and eigenvectors rearranges the dimensions in an n dimensional space so that all axes are perpendicular.

Appendix IV
Environmental Chain
of Custody

Client:	Contact Name:	Phone:	Fax:
Address:	City:	State:	Zip:
Purchase Order #:	Project Name/No.:	E-mail:	
Sampler (Print/Sign):	Copies To:		

Lab Use Only	Project Supervisor:
Shipping:	
Comments:	

Environmental Service Descriptions

Sample Description (Sample identification and/or site number)	Date/Time Collected	Matrix	Temp	Nutrient Analysis	Metal Analysis	Total Coliforms	Hardness	Pesticide Analysis	Volatile Organics	Radioisotope	BOD	Other Biologicals	Chlorides	Oil and Grease	Suspended Solids	Turbidity	Other (specify)
1																	
2																	
3																	
4																	
5																	
6																	
7																	
8																	
9																	
10																	
11																	
12																	
13																	

Relinquished By:	Date/Time:	Received By:	Relinquished By:	Date/Time:	Received By:

Index

T

T - #0385 - 071024 - C4 - 234/156/14 - PB - 9780367386344 - Gloss Lamination